Advancing Global Bioethics

Volume 9

Series editors
Henk A.M.J. ten Have
Pittsburgh, USA

Bert Gordijn
Dublin, Ireland

The book series Global Bioethics provides a forum for normative analysis of a vast range of important new issues in bioethics from a truly global perspective and with a cross-cultural approach. The issues covered by the series include among other things sponsorship of research and education, scientific misconduct and research integrity, exploitation of research participants in resource-poor settings, brain drain and migration of healthcare workers, organ trafficking and transplant tourism, indigenous medicine, biodiversity, commodification of human tissue, benefit sharing, bio-industry and food, malnutrition and hunger, human rights, and climate change.

More information about this series at http://www.springer.com/series/10420

Joseph Tham • Chris Durante
Alberto García Gómez
Editors

Religious Perspectives on Social Responsibility in Health

Towards a Dialogical Approach

Springer

Editors
Joseph Tham
School of Bioethics
Ateneo Pontificio Regina Apostolorum
Rome, Italy

Chris Durante
Saint Peter's University
Jersey City, NJ, USA

Alberto García Gómez
UNESCO Chair in Bioethics
and Human Rights
Rome, Italy

ISSN 2212-652X ISSN 2212-6538 (electronic)
Advancing Global Bioethics
ISBN 978-3-030-10123-7 ISBN 978-3-319-71849-1 (eBook)
https://doi.org/10.1007/978-3-319-71849-1

Softcover re-print of the Hardcover 1st edition 2018

Printed on acid-free paper

This Springer imprint is published by the registered company Springer International Publishing AG part of Springer Nature.
The registered company address is: Gewerbestrasse 11, 6330 Cham, Switzerland

Religious Perspectives on Social Responsibility in Health: Towards a Dialogical Approach

United Nations Educational, Scientific and Cultural Organization

UNESCO Chair in Bioethics and Human Rights Roma, Italy

The UNESCO Chair in Bioethics and Human Rights (www.unescobiochair.org) organized a workshop on "Bioethics, Multiculturalism and Religion," and the papers submitted here form the collection of this edited volume.

This UNESCO Chair was established in 2009 with two Roman universities. It offers a university framework of reflection and study, providing information and fostering the application of bioethical principles in science, medicine and new technologies based on the Universal Declaration on Bioethics and Human Rights. Through integral education, research and information, it seeks to contribute towards the recognition and promotion of a global and integral vision of bioethics, bringing to light universal values and principles as well as social and legal implications in relation to human rights. By creating a community of persons who are interested in these values, it seeks to promote cultural dialogue and encounters in a spirit of solidarity.

Preface

UNESCO has played a leading role in bioethics within the UN system. Its unique multidimensional mandate in natural, social and human sciences, along with education, culture and human rights, places the organization in a strategic position to develop an ethical framework for bioethics and ethics of science and technology. Countries members of UNESCO have adopted the only three global normative instruments in bioethics, namely:

- *Universal Declaration on the Human Genome and Human Rights* (1997, endorsed by the UN General Assembly in 1998)
- *International Declaration on Human Genetic Data* (2003)
- *Universal Declaration on Bioethics and Human Rights* (2005)

In the field of bioethics, UNESCO has become, and for more than 20 years, the leading agency to develop not only a laboratory of ideas but also the mechanism to disseminate normative instruments and to implement them. It is still the case in many countries, and at a regional level, the lack of well-developed capacities to deal with bioethical issues; lack of expertise: centres and committees and the lack of appropriate legislation, guidelines and teaching programmes.

This is why UNESCO has been devoted to support countries to establish and reinforce their bioethical infrastructure through activities aimed at the institutional and the professional dimensions. The first one is by establishing national bodies (such as National Bioethics Committees) that can advise the government on bioethical issues. The second dimension is through education. The main projects are Assisting Bioethics Committees and Ethics Education Program. The most important features of the Ethics Education programme are the production of a unique global core curriculum in bioethics, the training of ethics teachers and the development of training material in bioethics for researchers, health-care personnel, judges, journalists, among others.

UNESCO has also fostered the development of academic reflection, research and teaching activities in partnership with universities and academic institutions, mainly through the establishment of Chairs. There are around ten UNESCO Chairs in Bioethics around the world. The UNESCO Chair established at the *Ateneo Pontificio*

Regina Apostolorum and *Università Europea di Roma* has been focusing, for the past 8 years, in advancing the bioethical reflection in the context of multiculturalism and religion, as well as the promotion of bioethics through art. In so doing, it has constantly helped the interpretation and implementation of the Universal Declaration on Bioethics and Human Rights by opening cross-cultural and cross-religious debates on different articles of the Declaration and diverse bioethical issues, allowing the exploration of different ways of applying the principles in different contexts. This is a crucial task, as bioethical issues call for an inclusive and pluralistic approach.

According to UNESCO's Constitution, the activities in promoting science and international co-operation should serve as a channel to address the basic problems and needs of the world population, such as poverty and social development. In fact, the Universal Declaration on Bioethics and Human Rights contains a global understanding of bioethics, including social and environmental dimensions, along with the biomedical and technological ones.

In the case of this particular workshop, Article 14 of the Declaration on Bioethics and Human Rights was chosen, an article that fully underlines the social dimension of bioethics and the wide spectrum of actors responsible for health. By choosing this article, UNESCO Chair in Bioethics at the Ateneo Pontificio Regina Apostolorum and Università Europea di Roma "has provided a necessary space for reflection to address the dilemmas of resource allocation, global health, and social responsibility" and "exploring the relatioship between Global Health and Social responsibility and their link to Global Bioethics" (García and Sotomayor 2018).

Moreover, Article 14, Social Responsibility and Health, is paradigmatic as it includes the social dimension of bioethics as a part of the principles of this domain. It is a unique contribution to conceptualize and institutionalize global bioethics. It was largely discussed both by the independent experts of the International Bioethics Committee, the diverse regional consultations held during the process of the elaboration of the Declaration and the members of the Intergovernmental Committee on Bioethics. The discussions around it clearly showed the different approaches to bioethics; hence its inclusion is a significant step for the international community. During the drafting process, some members of the group underlined that the "social responsibility that derives from bioethics, demands that society deals with crucial issues to find solutions based on criteria of equity and justice" (Martínez-Palomo (2009).

I would agree with Father Joseph Tham, in the introduction of this volume, that through the different papers and during the workshop, it was easy to perceive the reality of the debate, of the discourse and of the emotions in the interreligious dialogue, with his call for a committed search of the commonalities and differences and his appreciation for the deep and sincere expression of each tradition during the presentations and discussions.

Some of the many interesting topics addressed during the workshop and that the reader can enjoy in detail in each of the chapters are as follows:

> Discussions around the definition of health: What is it? What are its different dimensions? How to embrace a wide definition and avoid at the same time a vague concept.
>
> Some of the religious perspectives discussed are as follows: "Who is the other?" To whom is the obligation to provide with health care, that is, to whom are we obliged, both the national governments and the international community? What is the difference between moral and legal obligation and charity and solidarity, particularly in social responsibility and heath care, to humanity? To national societies? To families? Does this social responsibility be thought as a legal or as a moral obligation? To everybody at the same level? How is minimum defined and how is it distributed? Which is the best approach to the right of health? International? Egalitarian? Both? Realistic, feasible vs idealistic? Is this a question of equality of equality of what?

There are very interesting considerations about the role of faith in ethics and law; in a secular and globalized context, human rights is the framework of the UN system in general, and the UNESCO Universal Declaration on Bioethics and Human Rights is the first global normative document that specifically links bioethics and human rights. Hence, one of the very interesting aspects of some of the papers is that they allow to continue and to enrich the dialogue and reflection between human rights and religion. Already the consultations held during the process of drafting the Declaration involved meetings and inputs from leaders of many religious traditions. However, as mentioned above, this dialogue and reflection about the implementation of the principles, and their links with human rights, is a living process.

In hand with the authors, the reader will have the opportunity to dig deeper into the differences and commonalities among various religious perspectives, moreover and very interestingly interreligious differences. In some cases more than in others, "tensions arising from the contemporary versions of a religious tradition while confronting modernity, between religious credence and pragmatic concerns, between theological and juridical approaches" (Tham 2018).

Learning about the differences within the same religion can not only be enriching, but as Guessous states: "the debate must continue among and between cultures and religions. Involving religious scholars and thinkers is necessary and beneficial for a democratic discussion and for the promotion of the values of Human Dignity, Equality, Justice, Autonomy, Responsibility and Solidarity for the wellbeing and peace of all Humans" (Guesssous Nouzha 2018).

I wish you all an interesting reading.

UNESCO
Paris, France

Dafna Feinholz

References

García, A.R., and C. Sotomayor. 2018. Chapter 20: Convergence and cooperation in social responsibility in Health. In *Religious perspectives on social responsibility in health*, ed. Joseph Tham, Chris Durante, Alberto García, 177–186. Cham: Springer.

Guessous, Nouzha. 2018. Chapter 15: Social responsibility in health. A contextualized lecture of Islamic spirit and cultures. In *Religious perspectives on social responsibility in health*, ed. Joseph Tham, Chris Durante, Alberto García, 133–140. Cham: Springer.

Martínez-Palomo, A. 2009. Article 14: Social Responsibility and Health. In *The UNESCO Declaration on Bioethics and Human Rights. Background, principles and application*, ed. Ten Have H, Jean M.S, 219–230. Paris: UNESCO Publishing. For a full description of the development of all the principles of the Declaration, refer to this publication.

Tham, J. 2018. Introduction. In *Religious perspectives on social responsibility in health*, ed. Joseph Tham, Chris Durante, Alberto García, 1–10. Cham: Springer.

Contents

About the Authors

Peter Au-Yeung is the chairman of the Diocesan Bioethics Committee, Catholic Diocese of Hong Kong as well as a specialist anaesthetist working in a public hospital in that city. He has written chapters in the book *Medical Law and Ethics* in Hong Kong and taught in a diploma in Catholic bioethics course given in association with the local Catholic diocese. He also lectured in various Congresses of the Asian and World Federations of Catholic Medical Associations.

Roland Chia is a Chew Hock Hin professor of Christian doctrine at Trinity Theological College and theological and research advisor of the ETHOS Institute for Public Christianity, Singapore. He is author and editor of many books including *The Ethics of Human Organ Trading* (2009) and *Hybrids, Cybrids and Chimeras: The Ethics of Interspecies Research* (2011).

María Elizabeth de los Ríos holds a Bachelor of Arts in philosophy at Universidad Iberoamericana in Mexico City and graduated cum laude. She also graduated with cum laude from a Master of Science in bioethics at Universidad Anáhuac in Mexico City. She also holds a PhD in philosophy at Universidad Iberoamericana and graduated cum laude. Her areas of interest are multiculturalism, postmodernism, violence and migration. She has worked at different Universities at Mexico City, and she is currently a professor at Universidad Anáhuac and Universidad Panamericana.

Chris Durante is an assistant professor in the Department of Theology at Saint Peter's University in NJ as well as a fellow of the UNESCO Chair in Bioethics and Human Rights, where he serves as the academic coordinator of the Bioethics, Multiculturalism and Religion workshops. Dr. Durante holds a PhD from the Faculty of Religious Studies at McGill University, a MA in religious studies from Georgia State University, a MSc in philosophy of mental disorder from King's College London and a BA in philosophy from Fordham University. His primary research and teaching interests are in religion, ethics and society broadly construed to include bioethics, comparative religious ethics, religion and politics, moral theology and moral philosophy. His publications have appeared in a number of academic journals such as the *American Journal of Bioethics*; the *Journal of Medical Ethics, Medicine, Healthcare and Philosophy*; the *Journal of Global Ethics*; the *Journal of Religious Ethics*; and the *Journal of Church and State*.

Ruiping Fan is Chair Professor of Bioethics and Public Policy in the College of Liberal Arts and Social Sciences of the City University of Hong Kong. He serves as Associate Editor of the *Journal of Medicine & Philosophy*, Associate Editor of the *Chinese Medical Ethics*, and Co-editor of the *International Journal of Chinese & Comparative Philosophy of Medicine*. He has published extensively on Confucian bioethics and comparative social and political philosophy.

Dafna Feinholz Mexican by birth, is UNESCO's Chief of Bioethics and Ethics of Science and Technology (Division of Youth, Ethics and Sports at the Social and Human Sciences Sector). A psychologist and bioethicist by training, she previously worked as a researcher in charge of the Department of Reproductive Epidemiology and was a member of the Mexican Research Council; she was the director of the Women and Health Program in Mexico, the academic co-ordinator of the Mexican National Commission of the Human Genome and the executive director of the Mexican National Commission of Bioethics, until she joined the UNESCO in 2009. From 2000 to 2006, she was the founder and chair of the Latin American Forum for Ethics Committees for Health Research (FLACEIS), an organization supported by the WHO. In her work for the UNESCO, she helps in setting up and supporting national ethics committees and ethics committee training around the world.

Colleen M. Gallagher has worked in the fields of social work and health-care ethics since 1985. Dr. Gallagher is currently the chief and executive director of the Section of Integrated Ethics in Cancer Care and an associate professor in the Department of Critical Care for the University of Texas MD Anderson Cancer Center where she has worked since 2008. Prior to this she served as a system director of Clinical Ethics for Resurrection Health Care in Chicago, Illinois, and before this worked for the Catholic Diocese of Cleveland and Catholic Charities Health and Human Services in Ohio. Throughout her career she has worked as a social worker and programme director for programmes serving the mentally ill and their families, persons with HIV/AIDS and the bereaved and other vulnerable populations. She received her doctoral degree in bioethics in health-care professions and social work from the Union Institute and University, Cincinnati, Ohio, in 2004. She completed an ethics fellowship in 1997 at Georgetown University, Washington, DC. She has also served in assisting the growth of both the social work and bioethics fields, having held several positions in professional organizations.

Alberto García Gómez is the director of the UNESCO Chair in Bioethics and Human Rights (Rome, Italy). He graduated law at Autonomous University Law School and is a Doctor of Law from Complutense University in Madrid. Presently, he is a professor of philosophy of law and international law at the School of Bioethics of Athenaeum Pontificium Regina Apostolorum in Rome. Furthermore he is a researcher of the Human Rights Institute at Complutense University. For 5 years he has been a member of the Steering Committee of Bioethics in the Council of Europe, and in 2005 he has been honoured with the National Prize of the Spanish Royal Academy of Doctors in the field of legal and social sciences.

Nouzha Guessous is a medical biologist and professor at Casablanca Hassan II University (Morocco). A researcher in bioethics and women's/human rights, she is associated to the Social Science Research Center, Jacques Berque Center for Humanities and Social Studies in Rabat. A former member and chair of the UNESCO IBC (2000–2007), she is currently a member of the ALECSO Science and Technology Ethics Committee. She is also a co-author of books and encyclopaedia as well as regular author of Newspapers *Tribunes on Bioethics and Human/Women's Rights*.

David Heyd is a Chaim Perelman professor of philosophy at the Hebrew University of Jerusalem. His main fields of interest are political philosophy, ethics and bioethics. He is the author of *Supererogation* (Cambridge, 1982), *Genethics* (Berkeley, 1992) and *Medical Ethics* (Tel Aviv, 1990) and the editor of *Toleration* (Princeton, 1996). Among his numerous articles, many are concerned with bioethical issues such as procreation, genetic technologies, euthanasia, sex selection, wrongful life, surrogacy and experimentation on human subjects. He served as a member of many ethical committees on surrogacy, on end-of-life issues and on ethics and genetics and the university's Helsinki committee.

Adina Halevy has a master's degree in social work and is a psychotherapist in private practice and a consultant to ICELP (International Center for the Enhancement of Learning Potential) in Jerusalem.

Jonathan Halevy is a physician and an associate clinical professor of internal medicine at the Hebrew University Medical School in Jerusalem, Israel, and a director general of Shaare Zedek Medical Center in Jerusalem.

Soraj Hongladarom is an associate professor of philosophy and director of the Center for Ethics of Science and Technology at Chulalongkorn University in Bangkok, Thailand. He has published books and articles on such diverse issues as bioethics, computer ethics and the roles that science and technology play in the culture of developing countries. His concern is mainly on how science and technology can be integrated into the lifeworld of the people in the so-called Third World countries and what kind of ethical considerations can be obtained from such relation. A large part of this question concerns how information technology is integrated in the lifeworld of the Thai people and especially how such integration is expressed in the use of information technology in education. He is the co-editor of *Information Technology Ethics: Cultural Perspectives*. His works have also appeared in *Bioethics*, *The Information Society*, *AI & Society*, *Philosophy in the Contemporary World* and *Social Epistemology*.

Stefan Kirchner His work is at the crossroads of human rights and the natural environment, with a particular focus on health issues and the sea. He works as a university researcher for Arctic Law at the Arctic Centre of the University of Lapland and *dosentti* (adjunct professor) for fundamental and human rights at the same university. In addition, he is sometime an associate professor (*docentas*) for the law of the sea at Vytautas Magnus University in Kaunas, Lithuania, where he also wrote his doctoral thesis on questions of Biolaw and the European Convention on Human Rights. Dr. Kirchner is admitted to the practice of law (*Rechtsanwalt*) in his native country Germany. In addition to his legal practice as an attorney, he served with Germany's Federal Maritime and Hydrographic Agency and has taught human rights and international law at universities in Germany, Lithuania, Ukraine and Italy, most recently as visiting professor for transitional justice at the University of Torino. Prior to his legal career, he worked as an emergency medical technician (*Rettungssanitäter*) for the German Red Cross.

John Lunstroth is a fellow of the UNESCO Chair in Bioethics and Human Rights (Rome, Italy). He is an ethicist in the broad, non-naturalist sense. He practised commercial law for about 20 years before returning to school in 2002, first for an LLM in health law and policy at the University of Houston Law Center (2004) and then for an MPH from Columbia University (2007). He has published on the regulation of alternative medicine, ethics of human experimentation, neuroethics, law of war, the theory of public international law and human rights and the intellectual history of the western tradition (focusing on law). He has taught health and human rights, public health law, history of the law of war and historicist legal theory. He has studied the history and philosophy of medicine (Chinese medicine, homoeopathy and scientific medicine) and science.

Vasantha Muthuswamy is a physician and retired as a senior deputy director general and chief of the Division of Basic Medical Sciences, Traditional Medicine and Bioethics and Division of Reproductive Health and Nutrition from Indian Council of Medical Research (ICMR), New Delhi. She was a WHO fellow at the Kennedy Institute for Ethics, Georgetown University. She is well recognized for bringing out the ICMR's "Ethical Guidelines for Biomedical Research Involving Human Subjects" in 2000 and the revised version "Ethical Guidelines for Research on Human Participants" in 2006. She was also responsible for the Guidance Document for Animal Experimentation, Guidelines for Stem Cell Research and Therapy, Guidelines for Safety Evaluation of Food Derived from GE Plants, Guidelines for Good Clinical Laboratory Practices and Guidelines for Safety Evaluation of Probiotics. She was also involved in the development of guidelines by WHO TDR and UNAIDS and national guidelines on research ethics at Nepal, Sri Lanka and Maldives. As the founder secretary of Forum for Ethics Review Committees in Asia-Pacific (FERCAP), she was a faculty member in bioethics workshops/conferences in more than 30 countries. She is currently the president of FERCI (Forum for

Ethics Review Committees in India) and chairperson of the Forum for Medical Ethics Society (FMES). She has received Lifetime Achievement Award from Indian Society for Clinical Research (ISCR), National Bioethics Conference (NBC) and FERCAP.

Aasim I. Padela is the director of the Program on Medicine and Religion and the Initiative on Islam and Medicine, an associate professor of medicine in the section of emergency medicine and a faculty member at the MacLean Center for Clinical Medical Ethics. He holds an MD from Weill Cornell Medical College, completed residency in emergency medicine at the University of Rochester and received an MS in health-care research from the University of Michigan. His Islamic studies expertise comes via a BS in classical Arabic from the University of Rochester, seminary studies during his secondary school years, and continued tutorials with several Islamic authorities. His research assesses how religion-related factors affect the health behaviours of American Muslim patients and the bioethical attitudes and medical practices of Muslim physicians. Dr. Padela also explores how scientific data can work in concert with Islamic moral reasoning and theology to develop a comprehensive, theologically rooted Islamic bioethics.

Claudia Ruiz Sotomayor Born in Mexico City, she holds an MD from Universidad Autonoma de Chihuahua, in Chihuahua, Mexico. She also graduated with a master's degree in bioethics from Anahuac University in Mexico City, and she graduated with a doctorate in bioethics from Loyola University in Chicago Il, USA. Currently, she is a fellow in clinical bioethics at MD Anderson Cancer Center in Houston, TX, USA. She has been a research scholar at UNESCO Chair in Bioethics and Human Rights since 2012 where she has worked in the area of multiculturalism, bioethics and religion.

Martha Tarasco received her medical degree from the Anahuac University in Mexico and obtained medical specialization in phoniatry in the National Institute of Human Communication in Mexico. She later received training in the UNAM Program for Clinical Medical Resident Researchers and sub-specialization in phoniatry in the University of Salamanca in Spain. Since 1986, she has worked privately in her medical office. She has worked in the Medical School of the Anahuac University coordinating the clinical cycles of students. In 1991, she organized a multireligious symposium on abortion for medical students. During 1992, she was trained in the Istituto di Bioetica of the Università del Sacro Cuore in Rome. In 1993, she opened the first master's degree of bioethics in Mexico. In 2002 she was the first executive director of the newly founded School of Bioethics at the Anahuac University, where she now works as a professor and researcher and leads the Institute of Humanism in Health Sciences. In 2005 she obtained her doctoral degree with summa cum laude at the University of Santiago de Compostela, Spain. She has published eight books, 27 chapters in different books and 37 articles in different journals.

Joseph Tham was born in Hong Kong and immigrated to Canada at the age of fifteen. At the University of Toronto, he first majored in mathematical sciences and then graduated from medical school. After several years of work as a family physician, he entered the seminary of the Legionaries of Christ and was ordained a priest in 2004. As a part of this preparation, he has obtained his degrees in philosophy and theology at Rome's *Regina Apostolorum* Pontifical university, where he also completed his postgraduate studies in bioethics. He successfully defended his doctoral dissertation with high honours on "The Secularization of Bioethics—A Critical History" under the direction of Dr. Edmund Pellegrino, former chairman of the President's Council on Bioethics. He presently teaches bioethics in *Regina Apostolorum* and was the former dean of the School of Bioethics. He is the editorial coordinator of *Studia Bioethica* and a fellow of the UNESCO Chair in Bioethics and Human Rights.

Samuel Weingerz Mehl is a surgeon who graduated from the UNAM University (Mexico City) with a specialty in higher studies of gynaecology and obstetrics based at the *Hospital Español* with master's and doctoral studies at the University Anahuac Faculty of Bioethics and has also played his private practice as a gynaecologist to the present time positions as chief of Division of Bioethics in *Hospital General Dr. Manuel Gea González* from 2002 to 2011 and director of planning and academic development in the National Commission of Bioethics from 2011 to February 2013. He is currently an external member of ethics committees in research and hospital of bioethics in *Hospital General Dr. Manuel Gea González* and is professor of ethics and bioethics at the undergraduate level in the faculties of humanities and health sciences at the Universidad Anahuac in Mexico City.

Ellen Y. Zhang holds a PhD in philosophy of religion from Rice University (Houston, TX). She is now an associate professor in the Department of Religion and Philosophy at Hong Kong Baptist University. She is also a research fellow at the Center for Applied Ethics at HKBU. Her research projects are related to Chinese philosophy, ethics and comparative studies. Her publications in ethics include "*'Weapons Are Nothing but Ominous Instruments': The Daodejing's View on War and Peace*" (2012), "*The Ethic of Givenness: An Ethical Dimension of Jean-Luc Marion's Theology of Gift and the Buddhist Principle of Dana*" (2011), "*Community, the Common Good, and Public Healthcare: Confucianism and its Relevance to Contemporary China*" (2010), "*What Is Personhood? Kant and Huayan Buddhism*" (2010), "*The Unity of Corporeality and Spirituality: Love in Daoism*" (2012) and "*The Neo-Confucian Concept of Body and Its Ethical Sensibilities*" (1999). She is now the editor-in-chief for *International Journal of Chinese and Comparative Philosophy of Medicine*. She is the member of the APA (American Philosophical Association), the AAR (American Academy of Religion) and the APCA (Association of Chinese Philosophers in America).

Chapter 1
Introduction

Joseph Tham

Abstract This volume is a collection of papers commissioned and presented in a 4-day workshop where bioethics experts from six major world religions—Buddhism, Christianity, Confucianism, Hinduism, Islam, and Judaism—gathered to discuss the meaning and implications of the social and ethical implications of the notion of social responsibility in their respective traditions and the ways in which their respective traditions could, or could not, uphold the tenets of Article 14. Judaism conceives social responsibility as universal equality, Islam as duty to God, Christianity as neighborly love, Confucianism as duty toward the family, Hinduism as a right balance of different duties, and Buddhism as active compassion. Each religion emphasizes a particular aspect of the principle. The summation of the different perspectives further enriches and completes the vision of social responsibility and promotion of health. Common concerns raised by these religious traditions are further explored. They regard the relationship between state and religion in an ever-secularized world, the relationship between health and salvation, the question of equalitarian distribution, and the relation between responsibility and rights.

Keywords Social responsibility · UNESCO Declaration on Bioethics and Human Rights · World religions · Multiculturalism

Our fourth encounter of this project of multiculturalism, religions, and bioethics took place in Mexico City from November 10–12, 2014. The theme for this encounter was to analyze Article 14 of the UNESCO Declaration on Bioethics and Human Rights concerning "social responsibility and health." This was part of an ongoing project of interreligious dialogue on the implications of the declaration, which resulted in the 2011 Rome workshop on Article 8 "Respect for Personal Integrity and Human Vulnerability" and the 2013 Hong Kong workshop on Article 12 "Respect for Cultural Diversity and Pluralism." Previous and pending publications by Springer are concrete fruits of these encounters (Tham et al. 2017, 2014).

J. Tham (✉)
School of Bioethics, Ateneo Pontificio Regina Apostolorum, Rome, Italy
e-mail: jtham@legionaries.org

J. Tham et al. (eds.), *Religious Perspectives on Social Responsibility in Health*, Advancing Global Bioethics 9, https://doi.org/10.1007/978-3-319-71849-1_1

The Mexico workshop has followed similar methodology of preceding ones, with authors assigned to write main papers from the perspectives of Buddhism, Christianity, Confucianism, Hinduism, Islam, and Judaism, alongside responses, one within the same faith tradition and the other from a different one. Each of the contributors was asked to explore Art. 14§1 and §2a of the declaration, with comments built upon previous encounters and focusing on the issues of justice, access, and social responsibility in healthcare according to the different aforementioned religious perspectives. Furthermore, they were asked to delineate how religion has contributed in improving the health of the society.

Article 14 of the UNESCO declaration is rather ambitious. It states in full:

1. The promotion of health and social development for their people is a central purpose of governments that all sectors of society share.
2. Taking into account that the enjoyment of the highest attainable standard of health is one of the fundamental rights of every human being without distinction of race, religion, political belief, economic, or social condition, progress in science and technology should advance:
   (a) Access to quality healthcare and essential medicines, especially for the health of women and children, because health is essential to life itself and must be considered to be a social and human good
   (b) Access to adequate nutrition and water
   (c) Improvement of living conditions and the environment
   (d) Elimination of the marginalization and the exclusion of persons on the basis of any grounds
   (e) Reduction of poverty and illiteracy

Our workshop purposefully limited the analysis to §1 and §2a of this article, concentrating on the more direct and proper aspects of health. This is not to ignore the fact that other social determinants mentioned in §2b–d are not irrelevant in their impact on a population's health (Baum 2016). However, in certain bioethics circles, the WHO redefinition of health as "a state of complete physical, mental and social well-being and not merely the absence of disease or infirmity" is considered too broad and probably unrealistic. Is it the scope of healthcare providers to offer nutrition, clean water, good education, and sound living environment to their patients? Moreover, by constantly widening the definition of health, there is the danger of medicalization—healthy lifestyles become commercialized and treated as marketable commodities, resulting in artificial creation of desires that are more products of social desires than curing diseases (e.g., cosmetic surgery, Viagra, medical tourism, etc.). On the other hand, there is an ongoing discussion whether one should widen the definition of health to include the spiritual dimension. This latter debate is especially pertinent in this book.

Interreligious dialogue is not always easy or pleasant. As witnessed in previous workshops, emotions can flare, and disagreements are not uncommon. This is good in the sense that we did not want superficial appreciation or testimonies of the different traditions. Yet, there is a need to seek truth through serious engagement without compromising one's dearly held beliefs or falling into relativism. In the previous volumes, I have discussed MacIntyre's approach of tradition-based conversation in the reality of global bioethics (Tham et al. 2014). We are also aware that this is not without polemics. For instance, Engelhardt is highly critical that such dialogues would be successful (Engelhardt 2006). During the workshop, Chris Durante offered a method of dialogue that was quite novel for our purpose. He proposes a three-stage method. First stage is "consensus building," where each religious tradition expounds their points of views and justifications without reproaches, while other interlocutors could enter the conversation to clarify the ideas and enrich the dialogue. It will be an ongoing process where there will be a continual search for sharable or "bridge" concepts. Here, relativism is avoided since one distinguishes truth from justification—one can possibly see the reasoning behind these justifications without assenting to their truth claims. The second stage consists in mutual acceptance of interpretive diversity when there is sufficient unanimity for a certain claim to be considered universal, while universality does not imply total uniformity but allows for "indigenous pluralism." The third stage consists in agreeing to disagree when the different positions are incommensurable. In this case, tolerance among the interlocutors is called for as a last resort. However, tolerance is not considered a principle but set up as parameters of permissibility distinct from endorsement or support (Durante 2015).

We will now briefly summarize the different papers presented in this book. The principles enshrined in the declaration are akin to seeds falling upon different terrains of human culture. The seeds flourish distinctively, growing into various plants and producing multifarious fruits. The fruits may have different shapes and flavors but are nonetheless similar, because the seeds from which they arose are the same—our common humanity. At the same time, the diverse terrains can have both limitations and peculiar richness for the fruit they produce. In this volume, we will see how the principle of social responsibility has engendered exceptional fruits in the terrain of religions: Judaism conceives it as universal equality, Islam as duty to God, Christianity as neighborly love, Confucianism as duty toward the family, Hinduism as a right balance of different duties, and Buddhism as active compassion. Each religion emphasizes a particular aspect of the principle. The summation of the different perspectives further enriches and completes the vision of social responsibility and promotion of health.

The Halevys presented a harmonious relationship in Israel between Jewish religious tenets and governmental policies. In the areas of healthcare, universal access is available to all its citizens which has a very high standard without any discrimination based on ethnic origins. Bioethical questions regarding infertility, informed consent, advanced directives, brain death, Good Samaritan law, and DNR orders at end of life are addressed according to the different voices of its multicultural constituents with special consideration from the Jewish angle. There is also pro-natality policy for infertility treatment in the allocation of resources and prioritization of cost-benefits in assuming advanced but expensive technologies. Mexican bioethicist Weingerz agrees and applauds these observations, stating that religion and policies are not in conflict but that the Jewish worldview can guide, orient, and inspire conduct in modern Israel in perfect conformity with standards of human rights espoused by the declaration.

In contrast, the two authors from the Islamic tradition of this volume disagree vehemently about what are the fonts of social responsibility. Padela, basing his sources from the theological and legal interpretation of the Qur'an and posterior writings, suggests that there are two possible analogues to social responsibility found, respectively, in *huquq al-ibad/Allah* and *Maslaha*. The primary goal of the Muslim believer is to worship God, and so these Islamic laws provide for the society and its members to fulfill this primary duty. Preservation of life, knowledge, property, lineage, and health are important interests of the state so that its members could worship in peace and tranquility. Hence, communal obligation to provide care derives from this standpoint. Moroccan bioethics scholar Guessous, in contrast, differs from Padela's approach by insisting that the principles enshrined in the UNESCO declaration are universal and applicable to all humans *qua* humans, and not specifically religious because life and health are so indispensable. She argues that there is not one interpretation of Islamic religious sources and bioethics with an Islamic outlook should embrace the universal human rights discourse, while religion complements it as a source of inspiration. Here, we see once again as in our former workshops the tension that contemporary Islam faces as it confronts modernity—between religious credence and pragmatic concerns and between theological and juridical approaches.

Most Christian contributors to this book agree that social responsibility and health have been a pivotal belief in Christianity now extended from the West to the rest of the world. The central belief that human persons are created in the image and likeness of God has had a great impact on the Christian ethical outlook. The person in need is our neighbor, another Christ who requires our aid. In serving God in our neighbor, we also are serving God. Thus, according to Kirchner, the Catholic natural law approach is in full harmony with the UNESCO Declaration of Bioethics and Human Rights. Au Yeung applies the ethical principle of proportionate-disproportionate/ordinary-extraordinary care to determine priorities of responsible care. Durante from an Orthodox Christian viewpoint wishes to emphasize the import of virtue formation in spiritual development or *theosis*, the process of divinization of the believer. Social responsibility at the personal level calls for the ability to recognize needs of the other. *Philanthropia*, or friendship to humanity, has been the Christian motivation for founding the first hospitals. Together with *diakonia*, a self-transforming virtue of service, practicing guest friendship and hospitality can bring about social transformation. Heyd, though a Jewish philosopher, agrees that healthcare provision seen as responsibility originated from the Good Samaritan account in the Gospel. He sees that Christianity has best answered the question of "Who is thy neighbor?" Heyd speaks of the history of supererogation that came from religious vows of poverty, chastity, and obedience made by religious orders of monks and nuns. These so-called evangelical counsels are above and beyond the normal duty of Christian living, which follows the example of the Good Samaritan who not only helped the neighbor in need but even provided for future to-be-incurred expenses. This idea of supererogation, to act beyond the call of duty, is closest to the ideal of treating every stranger with equality and universality in the current milieu of global ethics and human rights.

According to Fan, Confucianism would be open to Art. 14§1 in that governments have the duty to provide social services including healthcare to its citizens. He bases this from Mencius understanding of *tiandao* which mandates the benign ruler to promote and protect the well-being of its subjects. The flourishing of the family unit, the building block of society, requires stability which health provision would be a prerequisite. However, Fan finds the requirement of universalized healthcare as stated in Art. 14§2 somewhat dubious. According to Mencius, government should provide for its unfortunate members, but there must be a hierarchy of who to provide for with precise limits. The closest circles of relationships should receive aid and care before those who are further away. Thus, there would be a natural priority of care, starting out with immediate family members and expanding outward according to circles of proximity—extended family, relatives, neighbors, members in the same village, co-nationals, etc. Fan rejects a neoliberal approach in which he finds its radical egalitarian interpretation problematical. Chia also agrees about the role of government in human flourishing. Accordingly, the additional Confucian virtues of *yi*, *shu*, and *ren* could be supportive. *Yi* is sometimes translated to justice but has the meaning of being fair-minded: doing what is right and avoiding what is wrong. *Yi* is unremittingly context-related, applicable to different types of human relationships with a gradation of love and responsibility. *Shu* refers to reciprocity or putting oneself in another's shoes and thus the ability to feel compassion. The heart of compassion is *ren*, or benevolence which is the highest of all human virtues in Confucianism. In this system, care should theoretically be extended to all humanity, and its rejection of radical egalitarianism does not imply a rejection of equality of all humans.

On the word of Muthuswamy, Hindu theories on social responsibility are based on the balance between the demands of *karma* and *dharma* found in Bhagavad Gita. *Karma* theory proposes a somewhat resigned acceptance on individual social status, condition, and fate, whereas *dharma* inspires others to serve and alleviate the social needs of the less fortunate. Everyone has these duties of life toward the gods, ancient knowledge, ancestors, fellow humans, and the environment. These social responsibilities arise from spiritual discipline of self-perfection and virtuous living in order to attain liberation or *moksha*. Thus social responsibilities determine the general and special demands with a just balance of appropriate conduct in oneself, for the family, for the society, and for the nation. The 1946 secular Indian Constitution bestowed equality and fundamental human rights to all its citizens. Universal healthcare provision in India is compatible with this mandate, while recognizing that poverty and the caste system might mitigate this ideal. De Rios notes a contrast between the motivations of a *karma*-based fear of punishment in Hinduism with an imitation of the self-giving example of Jesus in Christianity. Lunstroth attests that in the West, which has undergone a separation of Church and state, there is still an ambiguity regarding who is truly responsible for the well-being of the society. He purports an interesting resolution of this tension between the secular Indian constitution and the expression of South Asian lifestyle. This is possible because he thinks the "western" idea of religion has been demystified in India, resulting in the identification of Hinduism and the state. This uneasy relationship between religion and the state will be subject of further discussion.

Buddhism takes on similar concepts like *karma* and *dharma* from Hinduism, with a major divergence vis-à-vis the annihilation of the self or *atman*. According to Zhang, a major obstacle to considerations of social responsibility is the Buddhist concept of *karma*. First, there is a possible problem that conceives *karma* as the law of causality and consequences, thus leaning toward a problematic ethic of retribution. Second, Buddhist emphasis on individual paths to liberation is not too concerned with socially engaged actions. Nonetheless, the notion of *karuna*, as compassion conjoined with wisdom, may offer a way out. Individuals require *karuna* to attain enlightenment. Thus, as a result of our interconnectedness, there arises a need for shared responsibility to help out the needy out of active sympathy. In this way, there will be personal responsibility implied in the works of charity, which is especially prominent in the Bodhisattvas tradition of involvement with assistance and healing. Hongladarom wishes to find a model compatible with justice, human rights, and social responsibility that are often deemed too individualistic for the Buddhist tradition. He postulates its foundation on a theoretical notion of justice in an ideal society where there is no competition or antagonism and will not be based on protection of individual rights. However, while this may work in a sublime harmonious world without any need to claim rights, it may not be practicable in the real world. In any event, such an attempt will shift the emphasis toward interconnected relations of individuals, whose *karmas* are intertwined in a common context, generating the needs to care for one another. In this conception of a collective *karma*, some movements are proposing a socially engaged Buddhism to counteract both the too otherworldly concerns of individual salvation and the individualistic emphasis on rights. These discussions in Buddhism further raise the association between individual and collective responsibilities. Rights and duties could perhaps be better understood in terms of solidarity, as a possible bridge concept between *karuna* and social responsibility.

We will now turn to the common concerns raised by these religious traditions which require further exploration. They regard the relationship between state and religion in an ever-secularized world, the relationship between health and salvation, the question of equalitarian distribution, and the relation between responsibility and rights.

A first concern deals with the relationship between the state and religion today for healthcare provision. In the past, healthcare is part and parcel of the domain of religious activities of charity and compassion. In the West, the first hospitals were monasteries, and the medical personnel were mostly priests or nuns. Due to a process of secularization, there has been gradual separation of the ambits of the church and the state that extends from government to the judiciary, education, and healthcare. Remnants of this can still be seen today where most educational and medical institutions are public but interspersed with a significant number that are founded with religious inspirations. The concomitance of public and religious health delivery systems varies a great deal among different nations and their religious makeup (Tham 2008).

The authors of this volume have given examples from their personal experiences. Israel, Morocco, Thailand, and India are mentioned as examples of coexistence of healthcare in the Jewish, Islamic, Buddhist, and Hindu contexts, respectively. In Israel, there appears to be a high standard of universal care in a multiethnic society inspired by Jewish ethical ideals. This is feasible due to the high living standard combined with small population. In Islam, the relationship between state and religion is an uneasy one, as theocracy meets modernity in many Muslim countries. Likewise, healthcare delivery would vary greatly from state to state, and Guessous' description of the Moroccan situation is an example of positive influence of Islam on social responsibility in health. Hongladarom describes an excellent system of sanitary provision in Thailand based on "the Buddhist teaching that no one can prosper or survive alone without the help that comes from all other beings." Muthuswamy reiterates statutory equality guaranteed by the Indian constitution, in a society which differentiation of people by caste is well ingrained.

Despite this apparent contradiction, there are surprisingly little protests or resistance to the differentiated availability of medical care in India, probably due to the resigned attitude toward one's *karma*. Even though the papers from Confucianism have not touched upon this subject, during the discussions, it came out how familism-based medical delivery could be achieved in countries such as China, Korea, Taiwan, Hong Kong, or Singapore. Communist China faces enormous challenges in this area, where paradoxically socialized medicine is not available or affordable to all. In contrast, Singapore offers a high standard and expensive healthcare system that allows extended family to pool their health insurance funds and is a possible example of Confucian family-based social model. It seems that health standards and availability are more dependent on economy and manageable population size, like Israel and Singapore, than religious practices or input. A different problem faces the secularized Christian West, where in general there are high standards of health. In spite of the fact that medical care arose from religious inspirations, there is a crisis of identity of these faith-based institutions. What value do religiously inspired hospitals add to patient care, when the main reason for their stay is to seek cure and not to be proselytized? Many of them are struggling to find the mission statement of their existence which consists in more than the presence of a chaplaincy department. Others face the prospect of closure or merger. In all these examples, we witness the evolving relationship between the state and religion at different crossroads of secularization and secularism. The strain between religious aspirations and the ideals of universal rights will continue under diverse aspects, as we see in the area of promotion of health and social development.

Another related theme regards the relationship between health in the physical sense and health in the spiritual or supernatural level in different religions. In this volume, the authors of Islam, Christianity, Hinduism, and Buddhism all emphasize the priority of spiritual well-being over physical ones. This precedence is exemplified in the Christian Gospel, "What good will it be for someone to gain the whole world, yet forfeit their soul? Or what can anyone give in exchange for their soul?"

(Matthew 16:26) In fact, the word health, derived from its Latin root *salus*, is used synonymously to mean salvation in its original etymology. It might sound strange to modern ears that in medieval times, a physician cannot treat a patient unless he or she has received the sacrament of confession beforehand. The rationale is that unless the patient has first been reconciled with God, any posterior physical healing would not be beneficial. Interrogatives on suffering and death, health and disease, and thus the whole enterprise of medicine cannot be easily separated from the religious answer. Certainly, in most religions, health and well-being are only a relative good, necessary condition for the spiritual practices to attain perfection or salvation.

In contrast, health for the contemporaries has become an absolute good. Previous lengthy dedication to build temples and churches is now replaced by the mad rush to establish hospitals and hospices. Physical *salus* (health) has usurped the place of eternal *salus* (salvation). Founding father of the Kennedy Institute of Ethics André Hellegers once said, "Medicine is increasingly being asked to provide heaven on earth." The divergence between the religious and irreligious vision regarding medicine and health is increasing. Medical science and technology have taken great strides in the past decades. Much of this is laudable for the flourishment of humanity and the common good. Yet, medical wonders can be put on a pedestal with a Promethean quest to cure all ills through technology. The technological imperative, to do everything technologically possible, can trump the values of human life, nature, and environment. The expenditure of human goods out of utilitarian concerns will require a constant vigilance and critique from religion to unveil the danger of hubris and, why not, idolatry (Tham 2014a, b).

Aside from the fact of hubris, there is concern that the quest for highest attainable standard of health of Art 14§2 is unattainable. Throughout the book, authors are dubious of this goal. Does the highest standard of health mean the best available care for all persons irrespectively of their place of origin? Would a country with a higher health standard be required to divest its resources to support developing countries, at the expense of their own citizens? Clearly, resources are not unlimited, and no government on earth can spend all their money on health alone. They each have to prioritize their needs and set certain limits—for instance, basic healthcare should be readily available before allotting funds for the latest treatment of rare diseases.

The ideal according to Art 14 infers that once the best medical service is apportioned for a particular population, its distribution should be fair and "without distinction of race, religion, political belief, economic or social condition." Clearly, this is an inheritance from modern liberal thinking beginning from Locke to the Rawls that frowns upon any kind of graded provision as discriminatory. Universal human rights means that once a society has decided on the cost of health, all its citizens are entitled to equal access. As the papers from different religious groups indicate, there seem to be a range of responses from some form of egalitarianism to prioritization based on tiers of closeness due to religious affiliation, nationality, tribalism, familism, or the caste system. The Judeo-Christian ethics seem most consonant with the ideal of universal healthcare, whereas other religions consider non-equal treatment a reality of the human condition. Islam and Buddhism are indifferent to social inequality because of their emphasis on one's equality before God or spiritual enlightenment. Hinduism recognizes the *karma* of unequal caste subjects as intrinsically predestined. The strongest objection comes from Confucian familism which advocates non-equal treatment to others. Confucians find the liberal concept of justice illusionary and inappropriate when unmoored from virtues such as *ren* and *yi* fostered in functional families.

The crux of the debate is related to the secularized underpinning of supererogation. When social responsibility is devoid of religious motivation, and supererogation is understood in terms of idealized universal rights where one can only opt-in or opt-out, any shades of preferential treatment will be considered unfair. What is missing in the human rights equation is the religious motivation behind helping my neighbor, this need to go beyond duty or supererogation. Universal rights presuppose an unflawed humanity that is able to rise above egotistic and selfish tendencies, to forego personal or national comforts in order to provide for the less fortunate. There is an overestimation of human virtue and generosity. As all religion recognize, there is a selfish and evil tendency within the human heart that has been repeatedly manifested in history, politics, social indifference, and moral decline. Catholic social teaching affirms this tendency:

> Man tends towards good, but he is also capable of evil. He can transcend his immediate interest and still remain bound to it. The social order will be all the more stable, the more it takes this fact into account and does not place in opposition personal interest and the interests of society as a whole, but rather seeks ways to bring them into fruitful harmony. In fact, where self-interest is violently suppressed, it is replaced by a burdensome system of bureaucratic control which dries up the wellsprings of initiative and creativity. When people think they possess the secret of a perfect social organization which makes evil impossible, they also think that they can use any means, including violence and deceit, in order to bring that organization into being. Politics then becomes a "secular religion" which operates under the illusion of creating paradise in this world. (John Paul II 1991, no. 25)

In light of this quote, we can better understand the tension between duties and rights noted in previous workshops and publications. The fact that religious traditions prefer the language of duties rather and rights is a humble recognition of the debility of the human condition. Egalitarian rights and munificent social responsibility might exist in the ideal world, product of the enlightenment. While there is a definite correlation between responsibility and rights, religious traditions challenges the fact that duties are derived from rights—Buddhism with its difficulty with foundation, Hinduism with its caste categories ingrained in *karma*, Confucianism with the priority of family and non-egalitarianism, Islam with its primary duty to God, and Christianity's critique of highest attainable health as an absolute good.

In addition, social responsibility would not be practicable without a sense of justice, fairness, previsions, charity, mercy, solidarity, compassion, and sacrifice, all of which are traditionally known faith-based virtues, be they Judeo-Christian, Buddhist, Confucian, or otherwise. The problem of just allocation of scarce medical resources will always remain no matter how advanced or rich a country has become. Members of society can learn again from the cardinal virtues of justice, fortitude, temperance, and prudence: justice and charity toward the needful, fortitude to carry on in the face of adversity and scarcity, temperance to humbly recognize the limitations of health, and prudence to wisely decide on the best use of limited resources.

Spiritual and mystical practices of different religions can help humanity overcome the temptation of technological hubris. There have been many models of heroes and saints who have provided examples of sacrifice and holiness in their exercise of medicine and nursing, like Damien of Molokai or Florence Nightingale. In our discussions, there is a certain dilemma about the disinterested nature of social responsibility. That is, our initial response to the needs of the Other, in the words of Jewish philosopher Levinas, should be *prima facie* because the needy patient confronts me personally. The response to offer assistance should not be based on one's personal gain, even as a duty to God or a means of spiritual enlightenment. While there is always a limit to how much good one can perform, the call to love our neighbor as oneself will always be there even in the most just societies. Social responsibility cannot be the task of the states alone but will require the motivation of its members to respond and reach out. Religion can help us reject the idolatry of well-being and perfection on the one hand while imbuing citizens the values to become sensitive to the needs of the less advantaged. In this way, social responsibility and solidarity can mystically become a form of worship.

References

Baum, F. 2016. *The new public health*. 4th ed. Oxford: Oxford University Press.

Durante, C. 2015. Dialogue despite diversity: Sharing norms when our moralities differ. *Studia Bioethica* 8 (1): 7–17.

Engelhardt, H.T., ed. 2006. *Global bioethics: The collapse of consensus*. Salem: M & M Scrivener Press.

John Paul II. 1991. *Centesimus annus*. Retrieved January 19, 2017, from http://w2.vatican.va/content/john-paul-ii/en/encyclicals/documents/hf_jp-ii_enc_01051991_centesimus-annus.html.

Tham, J. 2014a, May 13. Suffering technology (Part 1). *Studia Bioethica* 7 (1): 55–61.

———. 2014b, September 9. Suffering technology (Part 2). *Studia Bioethica* 7 (2): 50–56.

Tham, J., A. Garcia, and G. Miranda (eds.). 2014. *Religious perspectives on human vulnerability in bioethics*, 2014th edition. New York: Springer.

Tham, J., A. Garcia, and K.M. Kwan. 2017. *Religious perspectives on bioethics and human rights*. Cham: Springer.

Tham, J. 2008. The secularization of bioethics. *The National Catholic Bioethics Quarterly* 8 (3): 443–453.

Part I
Buddhism

Chapter 2
Karma, Compassion, and Dharma: A Buddhist Perspective on Social Responsibility and Healthcare—A Reflection in Light of Buddhism

Ellen Y. Zhang

Abstract This essay addresses social responsibility regarding the issue of healthcare from a Buddhist perspective. Traditionally, Buddhism is viewed as a contemplative religion which focuses on a solitary quest for emancipation and spiritual advancement rather than active engagement in the world or a social movement. The question now is how to bring Buddhism into conversation with the current debate on social responsibility and healthcare based on the *Universal Declaration of Human Rights* and the *Universal Declaration on Bioethics and Human Rights*. The discussion of the essay includes Buddhist concepts of karma, dharma, and compassion and the interpretation of these concepts to respond to relevant issues under the debate, particularly in terms of the Western ideas of human rights, equality, and social justice.

Keywords Buddhism · Healthcare · Karma · Dharma · Social responsibility

2.1 Introduction

Generally speaking, Buddhism is viewed as a contemplative religion which focuses on a solitary quest for emancipation and spiritual advancement rather than active engagement in the world or a social movement. Holding a tragic vision of the world with its inevitability of suffering, many of the early teachings of Buddhism, including the concept of karma (or the law of causality), intend to offer a positive attitude toward life in order to respond to the seemingly irreversible political, social, and

E. Y. Zhang (✉)
Department of Religion and Philosophy, Centre for Applied Ethics, Hong Kong Baptist University, Hong Kong, PRC
e-mail: eyzhang@hkbu.edu.hk

J. Tham et al. (eds.), *Religious Perspectives on Social Responsibility in Health*, Advancing Global Bioethics 9, https://doi.org/10.1007/978-3-319-71849-1_2

economic inequities and injustice of the caste system of the time.[1] As such, one may see a Stoic-like acceptance of the status quo on the one hand and the critique of social realities on the other hand in Buddhism. Then how do we understand Buddhist teachings today in light of the modern notion of social responsibility and its connection to the *Universal Declaration of Human Rights* (1948) and the *Universal Declaration on Bioethics and Human Rights* (2005), particularly in terms of justice and equal access to all in healthcare? There is no easy answer to this question since such concepts as human rights, equality, and justice are not derived directly from Buddhist doctrines. Nevertheless, it does not mean that the problem of incommensurability makes the dialogue between the two impossible.

2.2 Karma as Responsibility

Before we discuss the issues concerning healthcare, let us first look at the concept of social responsibility and how it can be appropriated from the perspective of Buddhism. Social responsibility, as it is understood today in the West, refers to an ethical duty indicating that an entity, be it an institution or individual, has an obligation to act to benefit society at large. However, responsibility for one's actions can also be understood as personal responsibility. Here, the boundaries among personal, social, and personal social responsibilities seem to be difficult to differentiate.

Then can we identify personal responsibility with social responsibility? I would like to go back to the Buddhist concept of karma from the Buddha's discourses. Very often we hear the claim that karma, according to our modern way of thinking, functions like a moral banking system, a credit and debit of good and bad. In this sense, karma carries the meaning that every act one performs has repercussions in the present moment which also reverberates into the future. That is, "[i]n its simplest form, the law of karma means that all actions, good or bad, produce their proper consequence in the life of individual who acts" (Chatterjee 1960, 74). This kind of interpretation remains contested because it seems to involve the notion of victim blaming or indifference to others' suffering in that the karma consequence (*vipaka*, *bao*報) aims at excusing or justifying suffering (including social suffering in terms of political oppression) as a necessary good. In fact, one of the criticisms of karma theory from a Western perspective (Christian in particular) is that it is immoral "because it is insufficiently sensitive to the human predicament, because it precludes underserved suffering, because it is a retributive ethical theory and no

[1] The example of Buddhist social responsibility is the welfare state created by the Buddhist King Ashoka (274–236B.C.E.). It is a popular view in Buddhist scholarship that Buddhism arose in India as a spiritual force against social injustices and inequality. It denounced India's caste system and advocated the equality of all men and emancipated woman, giving them complete spiritual freedom. Also, in *Aggañña Sutta*, the 27th *Digha Nikaya* (Collection of Long Discourses), there is a long passage where the Buddha criticizes the caste system. Nevertheless, Buddhism as a whole did not emerge as a social philosophy, neither a social religion in comparison to other religions like Christianity.

such theory is acceptable, because it gives us too much [personal] responsibility."[2] A complaint like this would not be a surprise given that responsibility for one's action or doing is seldom a clear-cut conclusion nowadays. For instance, when we ask, "Is the person responsible, or is society?" the answer we often get is that it is not either one or the other; rather, it is both or it is somewhere in between. This notion of "shared responsibilities" or "collective responsibilities" makes personal responsibility an interrelated issue and as such undermines the "personal dimension" in terms of moral responsibility. Such understanding of personal responsibility has also influenced the way Buddhism is perceived and interpreted in the West.

In his article, "What's Buddhist about Socially Engaged Buddhism," David R. Loy, a well-known Buddhist scholar, speaks of the non-duality of the personal and the social in Buddhist teachings. He contends that socially engaged Buddhism attempts to go beyond quietism and personal spirituality and is "perpetually engaged in a dynamic process of reflecting, interpreting, and transforming the socio-political and cultural situations in which they are embedded."[3] I have no doubt that one can support the idea of social responsibility from the Buddhist ideas of no-self, interdependent origination, compassion, etc., but those ideas do not by themselves make socially engaged Buddhism drop its emphasis on "personal responsibility" or "other worldliness" embedded in karma, a concept that is directly related to the belief of "rebirth." Therefore, in the *Upajjhatthana Sutta* (AN 5:57), the Buddha states:

> I am the *owner* of my karma. I inherit my karma. I am born of my karma. I am related to my karma. I live supported by my karma. Whatever karma I create, whether good or evil, that I shall inherit [emphasis by me].

To certain extent, Buddhism is a pragmatic teaching which starts from certain fundamental propositions about how a person experiences the world and how he/she acts and takes responsibility in it. As a law of causality, karma indicates the following meanings:

1. There is an action.
2. There is a causal action.
3. There is a causal action that points to a relation (i.e., another action).
4. There is a volitional or (free-)willed action [i.e., an action of free choice].

Here, the emphasis is put on an individual (the agent) action with his/her own volition/intention, and the action (as a necessary or sufficient condition) is considered a single act of a single individual or some function of more than one act of a single individual (e.g., the moral valence of the agent's actions to date).

However, due to the fact that an action can be (1) a single action by a group comprised of several individuals or (2) some function of more than one act by a group composed of several individuals, therefore, the individualness of an individual act

[2] See Karl Potter (2001 p. 231). Although the essay focuses on the Hindu tradition, the same thing can be applied to Buddhist teachings.

[3] Quoted from David Loy—Reproduction interdite (February/March 2004). http://www.zen-occidental.net/articles1/loy12-english.html (2/10/2015)

and responsibility becomes blurring. In other words, there is something that can be perceived as "group karma" or "collective karma." However, "collective karma" is a new concept, coined in recent years to fit the need to emphasize the social dimension in action, that is, each person is influenced as well as contributes to a *collective pool* of thought, emotion, and social conditions. To a certain degree, one can argue that individuals in a society share common institutions and belief systems which are the results of many different wills, both in the past and the present, rather than the consequence of any single individual action. Nevertheless, Shravasti Dhammika, a Theravadin monk, points out clearly, "Nothing like the idea of collective kamma [karma] is found in or even hinted at in the Buddha's teachings."[4]

In past decades, the term "social responsibility" became popular in Buddhism due to the emergence of what is called "socially engaged Buddhism" (which stems from "Humanistic Buddhism" 人間佛教 of the Taiwanese masters Taixu 太虛 and Yin Shun 印順 and was brought to the world by the Vietnamese master Thich Nhat Hanh), although Buddhism has a long history of taking care of the poor and vulnerable and the images of bodhisattvas in the Māhayāna tradition seem to reinforce the idea of compassion of social engagement. Engaged Buddhism emphasizes at least two aspects, (1) this worldliness and social responsibility and (2) collectivity and solidarity, both of which are formulated as a response to the critique from outsiders (i.e., to see Buddhism as being passive, quietist, otherworldly, and escapist). However, it would be misleading if we identify Buddhist notion of social responsibility completely with the concept used in the West today. After all, Buddhist soteriology focuses on self-help, self-responsibility, and the external dimension that is ultimately related to the internal.[5] Loy has recognized it when he insists that "Buddhism is thriving today because its basic principles remain just as true as when the Buddha taught them. If we try to find direct answers to our social problems in Asian Buddhist teachings, we will be disappointed, because those teachings were intended for people living in very different times and places."[6]

Whether personal or collective, the Buddhist theory of karma is fundamental to much of Buddhist doctrine, "because it provides a coherent model of the functioning of the world and its beings which, in turn, forms the doctrinal basis for the Buddhist explanations of the path of liberation from the world and its result, *nirvāna*" (Kragh 2006, 11). Nevertheless, Buddhism today has to address the social aspect of the karma that conditions the pattern of thought of behavior of an individual.

---

[4]Cited from http://buddhism.about.com/b/2014/01/20/collective-karma.htm (2/10/2015).

[5]Yet I am not suggesting that Buddhism cannot categorically develop a form of social teaching or movement that is similar to liberation theology (one stemming from Catholic roots) and the Social Gospel movement coming out of Protestantism at the turn of the century. Socially engaged Buddhism has done similar things in recent decades.

[6]See Loy, "What's Buddhist about Socially Engaged Buddhism."

2.3 Compassion and Social Responsibility

If we see karma as an idea emphasizing personal action and personal responsibility, compassion (*karunā* [慈]悲) then is an idea that focuses more on collective experience and shared responsibility, especially in Māhayāna Buddhism. For the Māhayāna tradition, compassion is linked in Buddhism as the supreme characteristic of a bodhisattva. According to early Buddhist teachings, to realize enlightenment, a person must develop two interrelated qualities: compassion and wisdom. Very often, the word compassion is thought to dwell on the negative aspect of human existence, sharing the suffering and pains of others or a willingness to bear the pain of others; this kind of compassion is sometimes translated as "pity," "mercy," or "sympathy" in English. But compassion also refers to "sympathetic joy," that is, sharing the happiness of others. In the first case, compassion is not merely a feeling of mercy or sympathy, but is understood to mean "active sympathy," that is, to do something with a helping hand. For a Buddhist, compassion is not a feeling of superiority; genuine compassion is about empowering others and helping them unlock strength and courage from within their lives in order to overcome their problems. Therefore, compassion is the central moral ideal that all the Buddha's followers are asked to practice toward all sentient beings everywhere and at all times.

Wisdom (*prajña* 慧) means "discernment" or "insight," the capacity to the world as it is (e.g., cultivating an awareness of interdependent origination, impermanence, and no-self). Wisdom also involves a capacity to use a "skillful means" *(upāya* 方便) to help others, which includes skills with regard to when and how to help. According to the Buddhist teaching, wisdom and compassion enable a person to have a deeper understanding of the conditioned co-production, seeing beings not as isolated beings but as patterns of relatedness. The images of bodhisattvas are meant to be a way to cultivate an awareness of a person's self-identification with the outer world and all beings. This view is taken by many socially engaged Buddhists who claim that no enlightenment can be complete as long as others remain trapped in a suffering life and that true wisdom is manifested in compassionate action.

Stories passed by in the Buddhist tradition tell us how Gautama Siddhartha, the Buddha, showed his concern for the health of his followers. He said to his disciples: "If you will not take care of each other, who else, I ask, will do so? Brethren, he who would wait on me, let him wait on the sick" (*Vinaya Mahavagga,* viii, 26). The Buddha also commented on what kind of person qualified to care for the sick: a good nurse or physician who is capable of prescribing the proper remedies for sickness. The Buddha further added that the good nurse is one who cares for the patient out of compassion and is capable of cheering, encouraging, and comforting the sick. The Buddha's teaching like this clearly shows the compassion and care needed for those who are physically sick. Loving kindness (*maitrī*, 慈), another key Buddhist concept, is seen as an integral part of Buddhist medical practice.[7] The doctrine

[7] Compassion (*karuna*), loving-kindness (*metta*), sympathetic (*muttita*), and equanimity (*upekkha*) are called "four noble states of mind" in early Buddhism.

behind it is the very idea that human beings have shared experience of suffering and vulnerability, and as such there is interconnectedness of everything both ontologically and ethically.

In the long history of Buddhism, in both Theravada and Māhāyana, it is well known that a Buddhist Sangha/community, as a social network, provides all kinds of services (including healthcare service) to lay people who need help. We see similar things in the Buddhist NGO like the Tzu Chi Buddhist Compassion Foundation (慈濟會) as Liaison Service in Taiwan and mainland China today.[8] The foundation embraces the idea of "global citizenship" and shared social responsibility and has been contributing to better social and community services, which include medical care and education around the world. In a way, the Tzu Chi Buddhist Compassion Foundation embodies the idea of socially engaged Buddhism that makes a difference to the secular counterparts.[9] However, it should be noted that the idea of social responsibility expressed in Buddhism is always associated with Buddhist teachings concerning compassion, sympathy, and loving-kindness rather than concepts such as rights, fairness, and social justice. The main goal of socially engaged Buddhism is to open a new path "connecting inner and outer transformation" as Loy puts it.[10] Thus, engaged Buddhism does not only speak of social responsibilities but also speak of social actions. "Social action" means many different actions that aim at improving the social condition of humanity, such as the act of charity practiced by the Tzu Chi Buddhist Compassion Foundation. The idea of social action can be

[8]Thirty Taiwanese established Tzu Chi charitable organization in 1996, by setting aside part of their daily housekeeping money. It has since grown into one of Taiwan's largest philanthropic organizations. It now claims to have ten million volunteers globally and support charitable work overseas through its branches in 38 countries.

[9]Another example of socially engaged Buddhism is the Tiep Hien Order (also known as "the Order of Interbeing") founded in Vietnam in 1964 during the war. It derives from the Linji school (臨濟宗) of the Zen/Chan tradition in China. The words "tiep" and "hien" have several meanings. "Tiep" means to be "in touch with" and "to continue." "Hien" means "to realize" and "to make it here and now." (See Thich Nhat Hanh, 1987, p. 11.) Other examples in Asia include the International Network of Engaged Buddhists based in Thailand (founded 1989) and the Sarvodaya ("Awakening of All") in Sri Lanka (starting in the 1950s). For more information on socially engaged Buddhist movements, see Queen (1996).

[10]In my previous essay on Buddhism and human rights, I have pointed out that since Buddhism speaks of a shared human experience, the issue that may bother Buddhism regarding the language of rights is The notion of individualism, or to be more exactly, an atomic notion of individualism (in Charles Taylor's words) embedded in the very idea of human rights. Although people talk about collective rights, that is, the rights of a particular group or community (such as children's rights, women's rights and gay rights), the idea of human rights, whether positive or negative, is based on free-standing individuals and individual autonomy. See Zhang (2014a). In his article "The Concept of Equality in the Theravada Buddhist Tradition," Padmasiri de Silva discusses the ongoing debate on Buddhism's concept of equality and human rights, and contends that the Buddhist way of looking at the issue is of equality different. He maintains that one should examine the concept of equality by looking at the reciprocal relationship of people in various social positions. See the essay in Zhang (2014b).

understood as a new understanding of Buddhist teachings that are not limited to solve the fundamental problem of how "I" experience the world and how "I" act in it but how "we" experience the world and how "we" act in it. Social responsibility, or social karma, understood in this way, is internal and external, individualistic and collective.[11]

2.4 *Dharma* and the State's Responsibility

The question now is how to relate the Buddhist teachings mentioned above to issues concerned with health and healthcare articulated by the *Universal Declaration of Bioethics and Human Rights* (2005, hereafter, UDB). In the UDB, Article 14:1 states that "the promotion of health and social development for their people is a central purpose of governments that all sectors of society share." It further goes in Article 14:2: "Taking into account that the enjoyment of the highest attainable standard of health is one of the fundamental rights of every human being without distinction of race, religion, political belief, economic or social condition, progress in science and technology should advance." In Article 15:1 on sharing of benefits, it states:

Benefits resulting from any scientific research and its applications should be shared with society as a whole and within the international community, in particular with developing countries. In giving effect to this principle, benefits may take any of the following forms:

(a) Special and sustainable assistance to, and acknowledgment of, the persons and groups that have taken part in the research
(b) Access to quality healthcare
(c) Provision of new diagnostic and therapeutic modalities or products stemming from research
(d) Support for health services

Here the access to healthcare and essential medicines is linked to an individual's economic rights and social justice. Perhaps, no one articulates this argument better than Norman Daniels, an American bioethicist, who connects a theory of justice to health and healthcare, arguing that healthcare (both preventive and acute) is special because health is special because it is required in order for an individual to have a fair opportunity for the goods of the society (Daniel 1985). It is quite obvious that for Daniels, justice to health and healthcare is understood in light of justice in terms of fair opportunity for everyone.[12] But who is in the position to take this social

[11] In fact, Dalai Lama has even adopted the term "universal responsibility" or "global responsibility" in order to emphasize a nonsectarian approach to world problems humanity is facing today. See Dalai Lama in his speech "Universal Responsibility and the Good Heart" in *Dharamsala* (Library of Tibetan Works and Archives), 1096.

[12] In his well-known book, *Just Health Care* (1985), Daniels' claim that all individuals are entitled to a decent minimum of healthcare services rests on two basic arguments: (1) a robust commitment

responsibility? For Daniels the answer is the state. That is to say, it is the state's responsibility to make sure that every citizen has an access to a fair opportunity for the benefit of medical research and healthcare.

We have to say that such a liberalist view with regard to "fairness" is totally absent in the Buddhist tradition, although I am wondering if the notion of the "dharma of a king" in Buddhism (as well as in Hinduism) can be translated into the "duty of a state or government" in contemporary situation. In Buddhism, *dharma* is a word with a polyvalent meaning, such as the "cosmic law and order" and the teaching of Buddha. Meanwhile, *dharma* also refers to the performance of duties, i.e., acts which were fit to be carried out according to their responsibilities. According to Buddhism, due to his compassion for the suffering, pain, and illness of his fellow human beings and animals, the Buddhist King Ashoka declared, "Let healing be brought to man and beast." Let us put aside the non-anthropocentric stance in King Ashoka's statement, the idea of healing for everyone suggests a moral obligation of a king as well as a notion of all-inclusive care for all. In the early Buddhist texts, we read a passage that describes the duties (*dharmas*) of a king that that invokes humane and compassionate social action:

> What is the duty of a Noble king? Depending on the *dharma*, honoring it, revering, cherishing it, one should establish guard, ward and protection according to *dharma* for one's own household, troops, nobles and vassals, for Brahmins and householders, town and country folk, ascetics and religious, for beasts and birds. Let no crime prevail in your kingdom, and to those who are in need, give property. (*Digha Nikaya* 26)[13]

There is an inclination to view King Ashoka's kingdom as a "welfare state" particularly when how the king was very much concerned with the creation of social conditions favorable to people's well-being during his reign.[14] According to recent studies on Ashoka's edicts, an official order issued by him in authority, King Ashoka did make a distinction between what he calls "state morality" and "private morality." The former was what he based his administration upon and

---

to equality of opportunity (construed in a Rawlsian manner) that recognizes that good health is especially important in realizing most species-typical life plans, and (2) healthcare services are specially strategic in achieving improved health status. As such, it goes along with the UDB's statement "the promotion of health and social development for their people is a central purpose of governments that all sectors of society share." In his later work, *Just Health: Meeting Health Needs Fairly* (2007), Daniels has toned down his original statement a little bit by saying that a principled account of justice in Rawls' work requires a new perspective on a fair process for setting limits to healthcare. Yet, his major argument remains the same. Here, Daniels identifies "health" with "healthcare," which he interprets as requiring universal access, and benefit-sharing as distributive justice.

[13] Damien Keown (2000, 2005) contends that the duty of the king in early Buddhism implies a notion of human rights.

[14] For instance, Walpola Rahula argues that "Buddhism arose in India as a spiritual force against social injustices, against degrading superstitious rites, ceremonies and sacrifices; it denounced the tyranny of the caste system and advocated the equality of all men; it emancipated woman and gave her complete spiritual freedom" (Rahula, 1978).

what he hoped would lead to a more just and more spiritually inclined society, while the latter was what he recommended and encouraged individuals to practice such as charity and helping those who were in need. Nevertheless, some scholars also have suggested that the reason for Ashoka to call for state morality is more politically motivated rather than following the teaching of Buddhism.[15] He even took the title *Devanampiya Piyadasi* which means "beloved of the Gods" or "he who looks on with affection." So it requires a careful examination to see if the Buddhist social teaching in particular historical context is relevant to the contemporary situation.

Nevertheless, Buddhists today support the idea of helping the poor and the sick for the sake of relieving pain and suffering, as what has been taught in Buddhism. They see such an act as a duty (a karmic act as well) for each individual. Buddhist compassion calls for sympathy, mercy, comfort, and almsgiving as shown in the Buddhist doctrines of *dāna* (gift-giving, the practice of generosity) and *punya* (doing good).[16] Hence, *dāna*, as an act of giving, is an encouraged social action for Buddhists. Nevertheless, *karunā* (compassion) is used more often than *dharma* (duty) when one speaks of "social actions" in terms of taking care of others, ranging from simple individual acts of charity and organized kinds of service, as well as "right livelihood" in and outside the helping professions through various kinds of community development.

In contemporary (and secular) society, when the king in the Buddhist tradition has been translated into the duty of an institution, i.e., government, and from a citizen's point of view then, it is much convenient to speak of a citizen's right than an institutional duty or virtue. Unfortunately, when we use the language of rights, things become more complicated. For instance, we understand that an individual's right to healthcare (e.g., "the highest attainable standard of health") does not exist unless there is a provider by either the government or a private insurance company. That is to say, the language of rights could be problematic since strictly speaking someone does not have a right to something unless there is a party or institution with an obligation to provide it.[17] In other words, someone does not have a right to something unless there is a person or an institution with an obligation to provide it. Does the Buddhist notion of duty or compassion support such concept of rights, as suggested by the UDB document? This question remains unclear in Buddhism, since Buddhism seems to suggest that there may be obligations, for instance, of a society

[15] For more information on this issue, see Ven. S. Dhammika's (1994).

[16] It should be pointed out that, from the early tradition, Buddhism does not see *dāna* as only material gifts, it can be spiritual gifts, that is, to offer the practice of precepts and wisdom to make better choices in life. This notion is connected to the Buddhist teaching of combining compassion with wisdom.

[17] But some people today would question if being committed to aiding the poor and vulnerable is equivalent to being committed to using government force to compel everyone in society to be charitable, that is, to make gift-giving obligatory. This question has not been an issue debated in Buddhist tradition.

to its members—for which there are no corresponding "rights." Thus a person may have an expectation, but not a claim.

As I submitted in my previous essay on Buddhism and human rights, Buddhism would accept a qualified notion of human rights as a protective mechanism, as Jay L. Garfield argues, "rights" are asserted when they are violated or threatened, and thus they are fundamentally protective and negative in character instead of a kind of entitlement (Garfield 1998, 1–14). At the same time, a Buddhist would argue that a person's success (or fortune) is also dependent upon the collective assets of the community, and thus without this "social capital" of the community, one cannot suppose that he/she has been able to make the fortune without having at the same time incurred a "debt," which means he/she has a social responsibility to pay back. But very often the moral reasoning behind the idea of "paying back" in Buddhism is done under the rubric of ethics of care and compassion instead of ethics of justice and autonomy. In this sense, I think the Buddhist view is closed to that made by the communitarian who prefers the talk on community and solidarity but not on a totalitarian state and an overcontrolling government.

2.5 The Sharing of Responsibilities and Benefits

Today, social responsibility requires social action that usually includes the act of giving (*dana*). For Buddhism in the past 2000 years, the major concern of the religion has been directed toward the goal of releasing one from the root causes of all existential sufferings, that is, greed, hatred, and delusion, all of which are also seen as the root causes of all social ills. However, the Buddhist practice (Theravāda and Māhayāna alike) throughout history has been dominated by an emphasis on an individual's quest for emancipation and spiritual freedom rather than active engagement in the world or a social movement. That being said, one can still find an intrinsic connection existent between the UDB notion of sharing of responsibility and benefits and the Buddhist recognition of the problem of greed, hatred, and delusion. The Buddhist teaching intends to offer a way for an individual to transcend his/her psychological cravings and excessive desires, especially if he or she is better off. Along this line of thinking, the idea of sharing of responsibility and benefits is similar to the Buddhist notions of charity and love-giving. As for the question whether or not Buddhism, like Marxism, favors in creating a social condition (i.e., an ideal condition) to promote true freedom for everyone and end suffering for each individual for the ultimate goal of self-transcendence (or self-actualization for the Marxist) remains another issue.[18]

[18] On several occasions, Dalai Lama claimed that he considered himself a Marxist. He said, "The economic system of Marxism is founded on moral principles, while capitalism is concerned only with gain and profitability. Marxism is concerned with the distribution of wealth on an equal basis ... as well as the fate of those who are underprivileged and in need, and [it] cares about the victims of minority-imposed exploitation. For those reasons, the system appeals to me, and it seems fair."

The UDB statement that "the promotion of health and social development for their people is a central purpose of governments that all sectors of society share" intends to call for an action that aims at actualizing a policy by which one can have "access to quality health care and essential medicines, especially for the health of women and children, because health is essential to life itself and must be considered to be a social and human good" (Article 14:2a). Buddhism would definitely side with the idea that health is essential to life itself and is also connected to a social and human good, yet it remains a question for the Buddhist if this social good is identified with the idea of "equal access." On the one hand, the Buddhist notion of interconnectedness points to a sense of "members one of another," and this arouses within us an awareness of the "we" and the "us" as opposed to an omnipresent "me" endorsed by radical individualism. Thus Buddhism would agree with the policy of maximizing the health of the least advantaged and the policy of priority to the sickest when deemed necessary. Yet on the other hand, Buddhism as a whole has seldom clearly stated that social and economic inequalities can *not* be allowed to exist. It should be noted that one can find a concept of equanimity (*upekkhā*平等心) in Buddhism referring to a state of mind that has developed a sense of the equality and interconnectedness of life, but it does not necessarily indicate a sense of distributive equality as we use the word in political philosophy today.

Therefore, Buddhism would agree with the idea of sharing responsibilities and benefit on the ground of human solidarity, but may not ask for a robust notion of "equality" and "fairness" on the ground of self-responsibility and practical wisdom, i.e., the capacity to make an informed and rational judgment in a specific situation. After all, Buddhism is a pragmatic teaching which starts from certain fundamental propositions about how we experience the world and how we act in it. The problem we have here is that although the UDB maintains that all human beings are "created equal" with respect to being equally entitled to respect and dignity, people are not created equally healthy since either natural or social "lottery" (or karma in view of Buddhism) gives some people good health and some catastrophic diseases. Therefore, neither the conception of equality nor the conception of rights can address the real problem of healthcare allocation and distribution. How do we define "adequate" and "decent"? How can we commit ourselves to cost controlling if we hold that there is an excess (or entitlement) to healthcare for everyone in the name of rights?[19]

Citation is from Ed Halliwell's article "Of Course, Dalai Lama's a Marxist" in *The Guardian*, June 20, 2011. Nevertheless, the Buddhist idea of freedom as minimizing non-volitionally caused suffering, whether in mind or body, is still not the same thing as what has been advocated in Marxism.

[19] Some ethicists in the West are skeptical about any argument that connects the language of rights and the principle of justice to healthcare. For instance, according to H. T. Engelhardt, Jr., the biggest problem of a state-controlled, single-tiered system, if one argues from the principle of utility, is that it can hardly maintain a healthcare system as it has usually promised, namely, (1) a decent equal care, (2) the same menu, (3) unlimited choices (for both patients and doctors), (4) cost controlling within manageable limits, and (5) no need for individual/family savings and fiscal responsibility (Engelhardt 2007, 27–45).

If we take the Buddhist concept karma seriously, we have to say that sharing of responsibilities does not exclude what constitutes individual accountability and personal responsibility. In other words, “social responsibility” should not be used as an excuse for someone not taking personal responsibilities. In the case of health and healthcare, it means that one needs to follow healthy lifestyles (e.g., exercising, weight controlling, and not smoking for people in developed countries) and be a good patient (keeping one’s appointments, heeding one’s physicians’ advice, and using a hospital emergency department only for emergencies, etc.).[20] But conditions that are beyond one’s personal control yet influence individual’s health significantly, such as clean water and air, should be, of course, the responsibility of the government and society. That is to say, individual responsibility/action and collective responsibility/action should go hand in hand.

Nevertheless, one can also argue that Buddhism may pay more attention to those “subjective” or volitionally caused karmic sufferings despite personal responsibility involved, because they are more likely to stimulate negative karmic action not only to those individuals but also to other people. I want to emphasize more personal responsibility here because we must acknowledge potential moral hazards entailed in universal healthcare when an individual fails in personal and fiscal responsibility. While supporting the idea of public healthcare for the purpose of promoting health and social development, we should also be aware of the potential healthcare morass caused by it when we use limited healthcare resources to satisfy virtually limitless individual needs. Despite that the state or government is important for safeguarding the effectiveness of the social good and public well-being particularly when a free market system fails, it does not necessarily and always function as a “moral authority” but as a “fallback mechanism” to counterbalance the limits and drawbacks of a free market.

2.6 Summary

The discussion in the essay intends to explicate the issue of social responsibility with connection to the *Universal Declaration of Human Rights* and at the same time tries to bring Buddhism into the contemporary discourse on healthcare and social responsibility. By looking at some key concepts in Buddhism, such as *karma*, compassion, and *dharma*, the paper intends to offer a direct response to specific articles in the UDB document in light of an interpretation of the core teachings of Buddhism in a contemporary context. We have spelt out the part in Buddhism which would support the basic spirit of the UDB while pointing out the part which may not go along with specific regulations of the UDB.

[20] I understand that in reality, it is difficult to implement such policies without dealing with difficult questions related to individual freedom and patients’ autonomy. It will be a problem anyone demands equal medical treatment from the government, yet no one welcomes coercive measures from the government to promote responsible behaviors.

After all, many questions remain further exploration: how should the modern Western concept of human rights be translated to non-Western religions such as Buddhism and to the local level such as in Chinese or other Asian societies? How can we compromise between rights and responsibilities or individualism and collectivism? It has a long-standing argument in terms of what constitutes "responsibilities" and what constitutes "rights." The *Universal Declaration of Bioethics and Human Rights* has surely raised many interesting and serious questions, and it is important for different religions and different ethical positions to look at those questions. At the end of the day, we may not get a final consensus, but at least we have been engaged in a critical dialogue wherein, with a hermeneutical "principle of charity," we acknowledge a cultural diversity while making an effort to understand one another.

References

Chatterjee, Satis C. 1960. *Fundamentals of Hinduism*. Calcutta: Das Gupta.

Dalai Lama. 1976. Universal Responsibility and the Good Heart. In *Dharamsala* (Library of Tibetan Works and Archives).

Daniels, Norman. 1985. *Just health care*. Cambridge: Cambridge University Press.

Daniels, Norman. 1998. Distributive justice and the use of summary measures of population health services. In *Summarizing population health: Directions for the development and population metrics*, ed. Institute of Medicine Committee on Summary Measures of Population Health. Washington, DC: National Academy Press.

Dhammika's, Ven. S. 1994. *The edicts of King Ashoka*. Berkeley: Dharma Net International.

Engelhardt, H.T., Jr. 2007. The family in transition and in authority: The impact of biotechnology. In *The family, medical decision-making and biotechnology, philosophy and medicine series*, ed. Shui Chuen Lee, vol. 91. New York: Springer.

Garfield, Jay L. 1998. Human rights and compassion: Towards a unified moral framework. In *Buddhism and human rights*. Richmond: Curzon Press.

Keown, Damien, et al., eds. 2000. *Buddhism and human rights*. Richmond: Curzon Press.

Keown, Damien. 2005. *Buddhist ethics: A very short introduction*. Oxford: Oxford University Press.

Kragh, Ulrich Timme. 2006. *Early Buddhist Theories of Action and Result: A Study of Karmaphalasambandha, Candrakirti's Prasannapada, verses 17.1–20*, Arbeitskreis für tibetische und buddhistische Studien, Universität Wien.

Loy, David R. "What's Buddhist about Socially Engaged Buddhism." http://www.zen-occidental.net/articles1/loy12-english.html.

Potter, Karl. 2001. How many karma theories are there? *Journal of Indian Philosophy* 20.

Queen, Christopher S. 1996. *Engaged Buddhism: Buddhist liberation movements in Asia*. Albany: State University of New York Press.

Rahula, Walpola. 1978. *Zen and the taming of the bull: Essays*. London: Gordon Fraser.

Thich Nhat Hanh. 1987. *Being peace*. Berkeley: Parallax Press.

Zhang, Ellen Y. 2014a. On human rights and freedom in bioethics: A philosophical inquiry in light of Buddhism. *Studia Bioethica* II.

Zhang, Ellen Y. 2014b. *Equality and the religious traditions of Asia*. London: Frances Pinter.

Chapter 3
Rights Can Coexist with Buddhism: A Buddhist Response to Ellen Zhang

Soraj Hongladarom

Abstract My response to Ellen Zhang's article in the book is that I argue that the conception of individual rights can coexist with the basic tenets of Buddhist philosophy that emphasize interdependence and compassion. Normally these two traits make it rather difficult to see how the individual rights, which are often regarded as presupposed by individuals as separate, autonomous entities, could be justified through a system of thought that emphasize just the opposite. However, if we imagine an ideal society where everyone acts as if they fully respect the rights of one another, we would see that the rights as expressly specified and written down are not necessary, as they are already implicit in the behavior of the members of the ideal society. According to the Buddhist viewpoint, such a society embodies the ideal of living together, where one is totally unselfish and is totally dedicated to the welfare of others. This society thus is morally perfect according to the Buddhist system of compassion and interdependence, and there is no need to emphasize the separate and autonomous nature of individuals in order to justify it. However, since the end result is the same, this ideal situation shows that it is possible that the individual rights regime can indeed coexist with Buddhism.

Keywords Buddhism · Rights · Individual · Interdependence · Compassion · Autonomy

Zhang's paper is an attempt to draw the view on social responsibility and rights from Buddhist teachings through an examination to two concepts, namely, karma and compassion. The idea is that the teachings of *karma* and compassion should be able to serve as a foundation for a conception of social responsibility and rights in such a way that an adequate and effective insight on health care could be found. Zhang's finding, however, is that Buddhist teaching leaves something to be desired

S. Hongladarom (✉)
Chulalongkorn University, Bangkok, Thailand
e-mail: hsoraj@chula.ac.th

J. Tham et al. (eds.), *Religious Perspectives on Social Responsibility in Health*, Advancing Global Bioethics 9, https://doi.org/10.1007/978-3-319-71849-1_3

in terms of providing an adequate and effective foundation. The reason, according to her, is that Buddhism focuses too much on the personal side of things and appears to neglect the social aspect of reality. Even though there are attempts, such as the Socially Engaged Buddhism Movement, or the movement in Taiwan which has attracted quite a lot of attention lately, Buddhism still retains a strong attachment toward the idea of personal responsibility toward salvation.

Zhang's view that there is an ambivalence in the way concepts of rights and social responsibility are founded in Buddhism stems from her belief that the concepts of rights and justice are not to be found directly within Buddhism. She says toward the beginning of her essay: "There is no easy answer to this question since such concepts as human rights, equality, and justice are not derived directly from Buddhist doctrines." Perhaps that is the case, but I would like to contend that Buddhism, being a rather flexible religion, does open up a space where interpreters can propose a view that core Buddhist teachings can actually provide an adequate foundation for social norms such as rights and justice. We can do this by having another look at the concepts of rights and justice. I would like to argue that there is a way to justify these concepts without having to rely on the idea of the autonomous individual. This idea is crucial for the liberal conception of rights and justice, as is well known. However, if we can propose a workable conception of rights and justice which is founded not on the standard view of the autonomous individual but on the view that individuals are constructs and are essentially social beings, then we can find a way to argue that the core teachings in Buddhist philosophy can indeed provide a firm foundation for rights and justice.[1]

Zhang's view that Buddhism has a difficulty in providing a firm foundation for rights and justice perhaps stems from her idea that rights have to be founded on the autonomous individual. Such a view is certainly widespread and indeed could be regarded as the mainstream view in western liberalism. According to this view, rights originate from individuals being in conflict with one another or more importantly from the need for ordinary individuals to protect themselves against possible abuses of power by the state. The antagonistic nature of the origin of the rights thus seems to be clear from the beginning. However, if we pay more attention on the *effect* of instituting the rights, then it is conceivable that the rights could also originate from another kind of environment where the antagonistic nature is much less emphasized. Imagine a society where everyone fully respects the rights of one another. In such a society, there does not seem to be any need for law nor its enforcement. Such society admittedly is ideal and almost never has a chance to be realized; however, the point is that such a society is conceivable, and a consequence is that individuals living in a society do not have to be always antagonistic against one another in order for there to be a fully developed system of rights. Individuals can be wholly supportive and helpful toward one another, and in this scenario their rights are certainly respected without the rights having to be written down in statutes. Here the effect, or the ideal situation that we would like to have, is one where

---

[1] For a clear introduction to Buddhist philosophy, see Siderits (2007) and Gethin (1998).

the rights are respected, but there are no hard and fast lines separating one individual from one another (such hard and fast rules would seem to be required if individuals are antagonistic toward one another). If we agree that a core tenet of Buddhist philosophy is centered around the idea that there is no hard and fast boundary separating one individual from another, then it appears that the core teachings are not incompatible with the existence of a rights regime. The ideal scenario that we are witnessing could well be described as one where there are rights and that those rights are fully respected by the individuals and enforced by the authorities, even if the rights might not be overtly recognized in statutes. In other words, there does not seem to be any conceptual conflict between a society that has a workable and effective list of rights (either in tacit or explicit forms) and one where individuals only derive their beings through their essentially social nature. Rights do not have to originate from antagonisms among individuals or from individuals being autonomous; on the contrary, the autonomy of the individual is based upon an effective conception and enforcement of the rights.

The core Buddhist metaphysical tenet, that of the emptiness of the individual, is thus not incompatible with the liberal ideal of autonomous individuals respecting the rights of one another. In a nutshell, to say that an individual person is "empty" is to say that his or her very being as an individual is devoid of any objectively inherent feature that would make her what she is. In Buddhism, all this is, ultimately speaking, an illusion, much like a rainbow is an illusion created by light and water droplets. We could say that what really exist are the light and the droplets. However, that does not mean that the rainbow does not exist. In the same vein, individual persons do exist in Buddhism, only that they can be analyzed as a composition of various factors that give rise to the appearance of there being an individual in the first place. The fact that there is nothing to the individual beyond these factors into which the individual can be analyzed (like the rainbow is composed of factors such as light and droplets) is what is meant by saying that the individual is "empty." Being empty in this sense thus is very different from being autonomous. Nonetheless the fact that an individual is empty does not preclude the possibility that he or she can and does possess a list of rights. These rights accrue to the individual by virtue of the ethical dimension that emerges from the fact that individuals have to live together and depend on one another for their survival. A key idea here is that individuals derive who they actually are from their relations with one another. This is the case only for the *social* nature of the individual, such as being a father entails there being a child or being a brother entails there being another sibling. But individuals do depend on one another also materially. I have to depend on a large number of others whose work ensures that I continue living. For example, the food that I buy from the market has to be transported from many places to the market, and the food has to be grown, processed, and distributed. All this requires a large number of people who work on each part of the process. Without these people working for me, the whole food production and distribution system would collapse, and I would be very hard pressed to stay alive. If the system of rights are there to ensure that a society survive and flourish, then the fact that people need one another for survival seems to entail that a system of rights is needed for the survival of human society too. The fact that individuals are either atomic and autonomous, or fluid and porous at the boundary as in Buddhism, is then secondary to this need for survival. If by conceiving of individuals as atomic and autonomous aids in carrying out the main objective of having a functional society, then such belief in atomic individuals is a useful one. But if by conceiving individuals to be fluid and fuzzy does have the same effect (i.e., a well-functioning society), then the belief that individuals are fluid and fuzzy is useful too. Individuals, in short, are not either autonomous or fluid in themselves. This way of looking at individuals corresponds much with the Buddhist tenet of the skillful means, or *upāya*, where the important matter is that the result is achieved, not the nature of the means that succeeds in bringing about the results.

Another of the Buddhist teachings that has given rise to a fair amount of debate and discussion is that of *karma*. Etymologically speaking, *karma* just means "action." A widespread misconception is that the law of *karma* is nothing but the law of fate. Everything is determined by the *karma*, and there is no room for personal freedom. However, that is not the case. What the law of *karma* says is that everything happens according to causes and condition—nothing happens by pure chance—and that there must be an adequate explanation to every occurrence. Thus the law of *karma* is very much like the principle of sufficient reason proposed by early modern philosophers such as Leibniz. It may be the case that when we make a decision, our decisions are influenced by factors leading up to the moment of the decision, but that does not mean that our decisions are completely determined by the prior factors in the sense that we are compelled by those factors that exist prior to the event in question. However, either to say that prior events always completely determine the event or our decision or to say that there are no determinations at all is a mistake. According to Nāgārjuna, the doctrine of emptiness says that prior events neither determine the decision at the moment nor is it the case that all events are completely separate from one another and no relations obtain among them. To say that prior events completely determine the action presupposes that the events are there objectively, and to say that events are not connected with one another at all would also be to presuppose that they exist objectively too, since they can exist without interacting with one another. According to the doctrine of emptiness, things are what they are only because they are links in the chain of causes and conditions, and what they are is only a matter of interdependence among these causes and conditions. Thus, according to Nāgārjuna, to say that our action is completely determined by prior events or to say that it is completely free is both a mistake, since our action is constituted by its relations and interdependence with other factors from the beginning. What this means is that the view that we are completely determined by our *karma*, or prior action, is false; we have certain leeways by which we could have acted otherwise. Thus we always have open before us a variety of choices even though what we actually decide could have in fact been influenced by prior factors. The view on free will and determinism in Buddhism is very complicated, and I cannot do justice with it in a short commentary here. Suffice it to say that in Buddhism we still retain our freedom of choice through our awareness of the choices facing us and our consciousness that the choice is freely taken. This implies that we still have to take responsibility for our own decisions, even though causes and conditions still could explain why we make the decisions that we do.

In her paper Zhang says that a problem with the law of *karma* is that it pays too much attention on personal responsibility. However, if the law is understood in the way I describe above, personal responsibility only plays a more limited role than commonly understood. A person is fully responsible for her action only if she has complete freedom in making the decision and in acting accordingly. But as her very being as a self, in itself, is not an objective entity existing entirely through itself, it cannot be said that she has complete control. Other factors, including the ones constituting her very self and her relations toward other objects, are also responsible. Thus *karma* is always a matter also of the environment. One of the Buddha's famous 38 auspicious acts given in *Mangala Sutta* says that to live in a place that is conducive to personal improvement is auspicious, as it is a factor that contributes to moving along the path culminating eventually in attaining liberation. Thus, to act in complete freedom is not enough; one has to place oneself in a good environment, such as an environment where there are good teachers and good spiritual friends who can help one with the practice. The idea is thus related to another point raised by Zhang concerning collective *karma*. The fact that living in a good place surrounded by good spiritual friends shows that the environment plays a role too. Buddhist teachings indeed recognize the importance of the environment, and the environment is certainly part of the collective *karma* that a number of people living in that environment share together.

We are now in a position where we can venture briefly on the connection of Buddhist teachings with the concepts of social responsibility and rights as they pertain to health care. Zhang's key question in the paper is as follows:

How do we understand Buddhist teaching today in light of the modern notion of social responsibility and its connection to the *Universal Declaration of Human Rights* (1948) and the *Universal Declaration on Bioethics and Human Rights* (2005), particularly in terms of justice and equal access to all in health care?

The question is a very crucial one in today's world. Anyone who would like to find a theoretical and conceptual connection between Buddhism and the real world has to find a good answer to it. What I can do is only to sketch a possible answer, which I attempted to do with the previous sections of this paper. How could Buddhist philosophy contribute to the justification of justice and equal access in health care? In other words, what does Buddhism have to say on the issue of social justice? Traditionally speaking, it is not much at all, as Buddhist monks seem more contented to isolate themselves from the commotion of the outside world in order to devote their time and effort to spiritual practice. What I have sketched, in a nutshell, is that Buddhism can in fact provide an adequate theoretical grounding of the concepts of rights and justice through the idea of interdependence of all things. We do not actually need an antagonistic or overtly legal conception of rights and justice in order to have a workable solution that just as the same or perhaps better. In other words, rights and justice do not have to be based on the idea that individuals have to separate themselves one from the other, always watching for potential conflicts and always defending oneself against all possible encroachment. There is another way of justifying rights and justice which is based more on solidarity of people rather than separateness, and this is much in the spirit of Buddhism. The rights and justice here do certainly include those that are directly relevant to health care.

To take a concrete example, Thailand is well known throughout the world for its effective health care allocation system where every Thai citizen and foreigner who registers to work in Thailand is almost completely covered. The government pays out to hospitals according to how many patients are registered with a particular hospital, and the hospitals have some freedom to manage the fund as they see appropriate. The system has worked quite well and has become more or less a model for other developing countries. The health care scheme here can be described in many ways. The official title for the scheme, literally translated from Thai, is the Thirty Baht for All Diseases Project, meaning, of course, that any Thai can have their diseases treated with only 30 baht (1 US dollar) in payment.[2] The justification of the project does not always mention the term "rights" or "justice," but often reasons are given in terms of the need for fellow citizens to help one another and to take care of each other. But the end result is all the same. No matter if one talks about the right of the patient to be treated adequately without paying too much (no more than 1 US dollar) or about the solidarity that Thai people share with one another, the end result is the same, namely, an effective health allocation scheme that takes care of more than 30 or 40 million Thai citizens. Right now the Thai government is required by law to implement the scheme, but even if they are not, it is very hard to imagine subsequent governments abolishing it because people have come to expect it.

The right to free health care has indeed been put in the current draft of the constitution which is being debated by the public pending the general referendum in July 2016, but one can imagine that even if the right were not there in the constitution, people have come to expect it anyway, and thus they feel they are entitled to it as if it were their legal right. And as for "social responsibility," the scheme shows that the government and the taxpayers who finance the scheme bear social responsibility together in that they are helping their fellow citizens prosper together, realizing, perhaps, the Buddhist teaching that no one can prosper or survive alone without the help that comes from all other beings.

References

Damrongplasit, K., and G.A. Melnick. 2009. Early results from Thailand's 30 Baht health reform: Something to smile about. *Health Affairs* 28 (3): 457–466.

Gethin, R. 1998. *The foundations of Buddhism*. Oxford: Oxford University Press.

Hughes, D., and S. Leethongdee. 2007. Universal coverage in the land of smiles: Lessons from Thailand's 30 baht health reforms. *Health Affairs* 26 (4): 999–1008.

Siderits, M. 2007. *Buddhism as philosophy: An introduction*. Aldershot: Ashgate.

[2]For a more detailed discussion of the 30 baht scheme, see, for example, Damrongplasit and Melnick (2009). For a more critical outlook, see Hughes and Leethongdee (2007).

Chapter 4
On Human Rights and Freedom in Biomedical Ethics: A Christian Response to Ellen Zhang

Colleen M. Gallagher

Abstract In the effort to find convergence of ideas and applications from various traditions, the discussion of human rights and freedom from many religious and cultural perspectives is needed. This response to Zhang seeks to find that convergence among the Buddhist and a broad Christian view. As Zhang presents some compatibility, this response seeks to add to those ideas by speaking about the roles of justice, compassion, and self-determination in the field of medical bioethics.

Keywords Human rights · Self-determination · Justice · Compassion

Applying religious teachings to legal concepts and determinations is always a challenge. Zhang has given us a means to consider Buddhism's teaching related to human dignity, rights, and freedom as stated in the Universal Declaration of Human Rights and the UNESCO Universal Declaration on Bioethics and Human Rights. She has pointed out that the concept of "human rights" is not directly addressed by Buddhism but is not incompatible with Buddhist tradition. Zhang approaches this by discussing the Universal Declaration of Human Rights (1948) which discussed "the rights of every individual" and how these are protected in legal form. She correctly points out that dignity, rights, and freedoms are often based in the idea of self-determination.

The challenge of addressing the distinction of self-determination and a tradition that teaches that human beings do not occupy a privileged position can be daunting. However, Zhang takes on the challenge and provides insight into ways of expanding on Buddhist teaching in ways that are not always evident to those from other traditions.

Self-determination seems most often to be approached from the perspective of the individual and the concerns of that individual in a legal format according to Zhang. While this assumption is not found in most religious traditions, it is in

---

C. M. Gallagher (✉)
Section of Integrated Ethics in Cancer Care, Department of Critical Care, University of Texas MD Anderson Cancer Center, Houston, TX, USA
e-mail: CMGallagher@mdanderson.org

J. Tham et al. (eds.), *Religious Perspectives on Social Responsibility in Health*, Advancing Global Bioethics 9, https://doi.org/10.1007/978-3-319-71849-1_4

Western legal traditions. Her discussion of philosophical terms is clear as she explains distinctions among dignity, freedom, and rights.

Zhang argues ways in which the Buddhist ethic of compassion could serve as a supplement to the concept of human rights. Hughes states that an obligation is to ensure that citizens do not fall into poverty as this begins a cycle in which other social ills follow. He likens this to a Buddhist approach to human rights and healthcare access (Hughes 2007).

One of the first challenges Zhang points out is that the concept of human rights and dignity might be seen as giving humans privilege in comparison to nonhumans. This itself would be different than Buddhist teaching.

While teachings may be quite different, Christian traditions teach that the individual lives in society and has duties toward others. Self-determination also must take into account the effects to others. Duties and rights are connected but not without distinction of each. Zhang points out that Keown discusses Buddhist teaching stating that duties and rights are mutually dependent. Zhang's discussion of the negative and positive rights is essential to understanding the distinction and the interplay among the two.

In Christian traditions the concept of social justice is more linked to human rights, while human dignity is more closely aligned with individual self-determination. This is not so different than what Zhang discusses regarding negative rights. A society may determine that it is a human right to have healthcare provided while at the same time not require a society to provide or pay for the same. The same may not remain constant when considering the role of the healthcare provider who takes on a duty through their profession. In other traditions the interplay of duties and rights is linked to the principle of justice.

Justice itself has several pathways. The most common are commutative justice, referring to that which is owed between individuals; legal justice, which refers to rights and responsibilities of being a citizen; contributive justice, which relates to what individuals provide to society; and distributive justice, relating to what society owes to its members.

Zhang discussed human freedom and the idea of self-determination and states that Buddhism does not have a direct teaching in relation to the modern concept of freedom. Her descriptions of Buddhist teachings on culpability and "right speech" are examples of how some ancient teachings can be applied in modernity without disruption.

Several of the current biomedical questions can be discussed in light of Buddhism and the discussion of human rights that Zhang provides. For example, the discussion of self-determination and human dignity can be linked to the concept of "the self" as done by Hongladarom in his work on privacy and genetic information. He considers providing genetic information as a convenient designation of self rather than as a property defining the human person. He links this to the doctrine of non-self. Similarly he discusses privacy in light of bioinformatics and respect for individuals (Hongladarom 2009).

Genetic information is specific to an individual but can be used by society when considering what is possible for groups of individuals with similar molecular genetic traits/markers. This is happening throughout the world today but with mixed thoughts about whether this use falls under common good or under rights of the individual. This is but one of the many challenges relating to biomedical ethics and the concepts being discussed. Balancing concepts of self, rights, responsibilities, social good, and justice relates to nearly every question regarding human action.

Globally bioethics is not specific to biomedical questions. It relates more closely to the idea of the role of individual in society and society serving needs of more than individuals. Again, justice is the principle most often used.

Zhang briefly mentions euthanasia and organ donation as examples but does not go into detail. In many Christian traditions, the question of personal freedom, rights of the individual to self and others, and the duty of individual providers come into conflict in these particular issues. These are both discussed by Masao Fuji using the Fourfold Holy Truths and the Noble Eightfold Path. In the discussion he uses the concept of compassion as in it the person is to respect his own life as well as the life of others (Fuji 1991).

Zhang gives a good discussion of balancing determinism and freedom as she considers the Buddhist concept of karma. In Christianity it is taught that a person is to be virtuous which requires volition of the individual. It is also taught that God is all-knowing but gives free will to humans. Ratanakul has written about the struggle for Buddhist precepts as Western bioethics has entered into Thailand. The approach used was the consideration of biomedical questions in light of Buddhist precepts that emphasize the motive of an action more than its consequence. Examples given are veracity and noninjury to life. Veracity is considered a duty. Noninjury to life is used to address questions of euthanasia. In addition, the link between the principle of justice and ideal of compassion is made. Using this Ratanakul is able to discuss social justice or what the UNESCO Universal Declaration on Bioethics and Human Rights would consider human rights (Ratanakul 1998).

Essential to any discussion of biomedical ethics and human rights is an understanding of how an individual or a community defines health. "According to Buddhism, to be active and healthy, one needs to live a spiritual life" (Hewaparthirane 2001 p. 6). Sorajjakool and Naewbood remind us of this as well as stating that "Disease, on the other hand, is the result of ignorance leading to craving desire, impurity, indulgence that leads to disharmony at the physical, emotional, and spiritual levels" (Sorajjakool and Naewbood 2010).

Among her most intriguing comments, Zhang uses what she calls the "negative golden rule." She quotes Samyutta-Nikaya "For a state that is not pleasant or delightful to me must be so to him also; and a state that is not pleasing or delightful to me, how could I inflict that upon another?" This manner of consideration of violation of what is called by others a human right is one that is likely understood and easily accepted by most religious, ethical, and legal scholars.

Zhang asserts that compassion, explained as active sympathy or willingness to bear the pain and sorrow of others, is a duty. If this is a duty that all can share, it can bring us to a common understanding of ethics (behavior) that also would be unwilling to inflict pain or sorrow. Is it possible that we might commonly agree that this alone would bring about the affirmation of human rights, dignity, and freedom?

References

Fuji, Masao. 1991. Buddhism and bioethics. In *Theological developments in bioethics: 1988–1990*, Bioethics yearbook, vol. 1, 61–68. Dordrecht: Kluwer.

Hewaparthirane, D. 2001. Buddhism, healing and wellbeing. *Wisdom 9* (1): 6–8.

Hongladarom, S. 2009. Privacy, the individual and genetic information: A Buddhist perspective. *Bioethics* 23: 403–412. https://doi.org/10.1111/j.1467-8519.2009.01716.x.

Hughes, J. 2007. Buddhist bioethics. In *Principles of health care ethics*, ed. R.E. Ashcroft, A. Dawson, H. Draper, and J.R. McMillan, 2nd ed. Chichester: Wiley. https://doi.org/10.1002/9780470510544.ch17.

Ratanakul, P. 1998. Bioethics in Thailand: The struggle for Buddhist solutions. *The Journal of Medicine and Philosophy* 13: 301–312.

Sorajjakool, S., and S. Nawwbood. 2010. Buddhism. In *World religions for healthcare professionals*, ed. S. Sorajjakool, M. Carr, and J. Nam. New York: Taylor & Francis.

Part II
Christianity

Chapter 5
Philantropic Healthcare: Christian Conceptions of Social Responsibility and Healthcare

Chris Durante

Abstract Drawing primarily from the Eastern Orthodox and Roman Catholic traditions, this essay will place the Christian tradition in dialogue with global bioethics by discussing the ways in which the ethical ideal of *philanthropia* speaks to the concept of social responsibility as it is explicated in the UNESCO Declaration on Bioethics and Human Rights (2005). In the Greek Orthodox tradition, God is described as being the great *Philanthropos*, and Jesus Christ, as a healer of body and soul, is described as the "great physician." These notions led early Christians to charitably tend to the needs of the sick as a central component of leading a Christian lifestyle. For this reason, *philanthropia* became part and parcel of *paideia*, or education, and the gold standard of Christian discipleship. It was under the auspices of *philanthropia*, or loving-kindness toward needy others, that hospitals and hospices were established as charitable institutions throughout Byzantium and later in the West as well. If we come to think of responsibility as a state of being that requires the cultivation of compassion and benevolence, which results in actively engaging in philanthropy and true physicianship, then it seems that actively fostering the individual's sense of social responsibility to provide healthcare as a heartfelt response to suffering can play a greater role in medical ethical education. In this way the problems associated with healthcare provision will not simply be reduced to their socioeconomic and political dimensions but can come to bear on our understandings of what it means to be a "good physician" in the moral sense. Hence, the issue of philanthropic healthcare can go beyond discussions of how we ought to redistribute material resources to the *redistribution of moral responsibility*, with both healthcare providers and those who educate them in medical ethics partaking in the processes of ensuring healthcare justice on both the local and global levels.

Keywords Philanthropy · Social responsibility · Healthcare · Global bioethics · Christian ethics

C. Durante (✉)
Department of Theology, Saint Peter's University, Jersey City, NJ, USA
e-mail: cdurante@saintpeters.edu

J. Tham et al. (eds.), *Religious Perspectives on Social Responsibility in Health*, Advancing Global Bioethics 9, https://doi.org/10.1007/978-3-319-71849-1_5

This chapter will address the concept of social responsibility in healthcare and global bioethics, as explicated in the UNESCO Declaration on Bioethics and Human Rights, in relation to traditional Christian moral and social thought. Beginning with an examination of the ways in which social responsibility is articulated and discussed within the Declaration, this study will then proceed to offer an analysis of the ethical concept of social responsibility from the perspective of traditional Christianity, by which I mean the Eastern Orthodox and Roman Catholic traditions of moral thought.

In 2005 the UNESCO Universal Declaration on Bioethics and Human Rights was issued as a form of soft law or a means of guiding global governance and influencing societal behaviors relating to bioethical issues. The drafters of the Declaration describe its scope as twofold, writing:

> 1. This Declaration addresses ethical issues related to medicine, life sciences and associated technologies as applied to human beings, taking into account their social, legal and environmental dimensions. 2. This Declaration is addressed to States. As appropriate and relevant, it also provides guidance to decision or practices of individuals, groups, communities, institutions and corporations, public and private (UNESCO 2005a).

As evident in the document itself, the principles and recommendations enunciated herein are intended to guide the legislative actions of States as well as the actions and behavioral dispositions of a wide array of collective bodies currently operative in what may be described as civil society. To this end, the drafters of the Declaration state that the first two aims of this Declaration are:

> (a) to provide a universal framework of principles and procedures to guide States in the formulation of their legislation, policies or other instruments in the field of bioethics; (b) to guide the actions of individuals, groups, communities, institutions and corporations, public and private … (UNESCO 2005b)

Hence, the Declaration arguably sees itself not simply as a form of soft law, which indeed it is, but also as a call for the cultivation of a civic ethics, on both local and global levels, that is attentive to the bioethical and health-related dimensions of social life. Consequently, it addresses both the political and legal and the civil and personal dimensions of moral life within the global public arena.

Going on to provide a rather robust conception of social responsibility with a wide purview of applicability, the Declaration delineates the parameters of its principle of social responsibility in Article 14, stating:

> 1. The promotion of health and social development for their people is a central purpose of governments that all sectors of society share. 2. Taking into account that the enjoyment of the highest attainable standard of health is one of the fundamental rights of every human being without distinction of race, religion, political belief, economic or social condition, progress in science and technology should advance: (a) access to quality health care and essential medicines, especially for the health of women and children, because health is essential to life itself and must be considered to be a social and human good; (b) access to adequate nutrition and water; (c) improvement of living conditions and the environment; (d)

> elimination of the marginalization and exclusion of persons on the basis of any grounds; (e) reduction of poverty and illiteracy (UNESCO 2005c).

Although legislative measures on a global scale are necessary in order to uphold UNESCO's principle of social responsibility and health, stated above, and realize its vision, law alone will not suffice. In accord with the sentiment expressed above in Article 14.1, the realization of this principle and the ends it seeks to achieve will require a transformation of personal and local communal dispositions that will produce forms of social agency capable of making such a vision a lived reality possible to ensure that we are not simply left with a set of inefficacious laws, which may or may not at times be upheld.

Drawing on ideas derived from Eastern Orthodox and Roman Catholic Christian moral and social thought, I will argue that social responsibility requires the recognition of both duties and virtues if it is to serve as a holistic concept and be an effective means of enacting widespread societal change. To be a *Christianoi*, or a Christian, is to be one who imitates the life of Christ, which speaks to the cultivation of character traits as well as the performance of actions. I will argue that a virtue-oriented approach to social responsibility must serve as a crucial supplement to the delineation of the deontological dimensions of the social responsibilities of both persons and organizations, especially those persons and organizations involved in medical education and the provision of healthcare.

The idea of "social responsibility" often carries with it notions of duties and obligations that individuals as well as communities hold toward the larger society. In this sense, social responsibility speaks to the ethical obligations of an individual person toward the community in which she is a member as well as the obligations that groups and communities have toward one another. The acquisition of virtues involve the cultivation of dispositions and the habituation of ethical behavior that leads to development of moral persons through a process of self-transformation in such a way that one's very sense of selfhood is intimately intertwined with one's moral identity as a person. Consequently, if one's sense of moral duty and obligation is to be sincere, it must arise from, and be related to, one's moral identity as a human person. From the perspective of Christianity, duty and virtue cannot be divorced from one another and reduced to distinct theoretical approaches to ethical analysis but are perceived to be integral aspects of the seamless cultivation of personal and social morality. As a result, authentic social transformation requires self-willed personal transformation on the part of individuals. As Patriarch Bartholomew of Constantinople has claimed:

> As we receive the message of Christ's commandment to love one another, we become the agents of change in the world. We become transformed, moving from glory to glory in a process of divinization (or theosis). Orthodox spirituality understands this process as one that is not merely individual but corporate. Society becomes transformed as well. We affect all whom we touch, by example and through our actions in the world (Patriarch Bartholomew 1997a, p. 64).

In regards to healthcare, the Christian tradition is ripe with examples of tending to the needs of the sick as a core feature of living an ethical life and embodying a commitment to the transfiguration of the world. A sense of responsibility to care for the health of others is evident in biblical passages, in which Jesus Christ told parables, such as that of the Good Samaritan, and in which He Himself acted as a healer of the sick and the embodiment of an unconditional benevolence toward the suffering of others. From the numerous saints who performed healing miracles and were involved in tending to the needs of the ill to the many Christian hospitals and healthcare centers operating throughout the world today, Christianity is a religious tradition with a deep-seated concern for social responsibility in healthcare that lays at the heart of its conception of the moral transformation of global society. Speaking from the Roman Catholic perspective, Pope John Paul II has stated that:

> The parable of the Good Samaritan belongs to the Gospel of suffering. For it indicates what the relationship of each of us must be towards our suffering neighbour. We are not allowed to 'pass by on the other side' indifferently; we must 'stop' beside him. *Everyone who stops besides the suffering of another person,* whatever form it may take, is a Good Samaritan. This stopping does not mean curiosity but availability (John Paul II 1984).

Personal responsibility is the ability to recognize and respond to the needs of others so that the focus is shifted away from the obligations one is bound to uphold toward the dispositions that an individual must cultivate in the process of becoming "available" to those who suffer in one's attempt to imitate and/or embody the beneficence of the divine as a core feature of their spiritual self-transformation. Patriarch Bartholomew has emphasized this point when he describes the Christian view of justice as a virtue, writing:

> In Holy Scripture, justice does not have only or principally the current meaning of the dispensation of justice, namely of justice being served. Justice carries a more extensive and comprehensive sense of virtue, such as expressed in the well-known aphorism of Aristotle: "Every virtue is contained in justice." The just person does more than merely comply with the law; the just person bears within oneself a higher conception of justice, namely of the perfect relationship of all things to one another. Thus, the just person is virtuous in every respect. For example, Holy Scripture describes as just whosoever shows mercy all the daylong and freely gives of oneself, without any legal obligation (Patriarch Bartholomew 1997b, p. 163).

As individual persons take such responsibilities upon themselves in the spirit of solidarity and fellowship, they too can become moral exemplars transforming the dispositions and affecting the decisions and actions of those around them so that as such a sense of responsibility permeates through our concentric circles of social relationality, entire communities can come to feel responsible not only for themselves and the weakest among them but also for the welfare and well-being of other communities as well. Christians will work in tandem with the State when its activities are just yet will not rely solely on State-based measures to ensure social justice and as their vision of social responsibility is not constrained by a State-oriented notion of politics but rather entails a deeper notion of the *polis* as an authentically interpersonal community.

The Christian idea that cultivating virtue speaks to an ethics of self-transformation and societal transfiguration will necessarily entail charity and love in any attempt at ensuring healthcare justice. The Latin idea of *Caritas*, charity, speaks directly to these issues with Thomas Aquinas having called it the most excellent virtue, which is capable of uniting one to God as it manifests itself as habitual altruistic love of neighbor. In the patristic tradition of Christianity, God is described as being the great *Philanthropos*, and Jesus Christ, as a healer of body and soul, is described as the "great physician." These notions led early Christians to charitably tend to the needs of the sick as a central component of leading a Christian lifestyle. Eventually, through the work of St. Basil this led to the creation of charitable hospitals that were themselves founded upon the Christian reconceptualization of the Hellenic idea of *philanthropia*, or being an "unconditional lover of humankind." Being adopted by early Christians as a means of describing God Himself, and finding its way into the Byzantine liturgical language, *philanthropia* came to represent a uniquely social and interpersonal dimension of Christian love that when embodied propels one to social action and ethical engagement with others – be they Christian or otherwise. As the Orthodox theologian and Church historian John McGuckin notes:

> Philanthrophy, thus, for Christians became exemplary of the perfection of love as manifested in the cosmos; something more public and social than *agape,* (which reflected chiefly the mutuality of charity among members of the Church), and something closely allied to the Church's duty to reflect in the world the impact of the philanthropy of God experienced within it that released if from bondage toward a new sense of compassion and love (McGuckin 2010a).

The Christian conception of ethico-spiritual life entails overcoming egoism and self-conceit through a transformation of the self into a lover of humanity, who will then actively manifest this love through compassionate and benevolent behavior toward everyone. In the Christian tradition, it is not simply enough to love God but also to actively love one's neighbor regardless of that person's social identity. For this reason philanthropy become an ideal that was part of *paideia,* or education, and became a gold standard of Christian discipleship. It was under the auspices of philanthropy, or loving-kindness toward needy others, that hospitals and hospices were established as charitable institutions throughout Byzantium and later in the West as well. The notion of hospitality, which stems from the ancient Greek concept of *xenia*, conveys the idea of "guest friendship" or the mutual respect between host and guest and the generosity of a host toward his guests. The notions of hospitality and hospital share the same conceptual origins and imply that an active willingness to serve others is held by she/he who is the host. This idea of service, or *diakonia*, again lies at the heart of the Christian message, becoming part and parcel of what it means to practice Christian discipleship, and, as I will demonstrate, plays an integral role in Christian vision of social responsibility as it relates to the practice of healthcare.

In its disambiguation, healthcare implies caring for and about health and, hence, involves a disposition that will propel the provider of such care to work toward the restoration of wellness so that she/he lacking health can continue to flourish as an individual person and member of the various concentric circles of community she/he belongs to. This was part of the "art of medicine" that involved a creative aspect that sought to restore health as a means of personal and communal flourishing rather than simply repair a physical "dysfunction" or correct a physical abnormality.

Coupled together, the empiricism of modern science and the financial corporatization of contemporary medical practice threaten to diminish the artistic and eudaimonic dimensions of healthcare delivery. The Christian outlook views the sick as suffering persons and not as mere consumers of medicine while it views healthcare providers as healers serving the health-related needs of the sick rather than being merchants of medicine. From the perspective of traditional Christianity, healthcare ought to be a philanthropic social force that seeks to bring about, restore, and preserve biological as well as personal vitality so that it may assist in cultivating human flourishing. Medicine is a practice whose *telos*, or aim, is healing, or to "make whole." From the Christian point of view, healing is not simply curing or treating but involves a restoration of the well-being of the person. As a practice, medicine is not simply an "art" or a "science" but an *askesis*, which comprises both. As an *askesis*, or practice, medicine or healthcare contains within itself sets of technical skills but also contains an implicit set of virtues pertaining to those persons engaged in the delivery of healthcare.

According to Catholic philosopher Alasdair MacIntyre, "the concept of a virtue … always requires for its application the acceptance of some prior account of certain features of social and moral life in terms of which it has to be defined and explained (MacIntyre 1984)." He goes on to assert that the first stage in defining a virtue is rooted in practices and defines a practice as:

> any coherent and complex form of socially established cooperative human activity through which goods internal to that form of activity are realized in the course of trying to achieve those standards of excellence which are appropriate to, and partially definitive of, that form of activity, with the result that human powers to achieve excellence, and human conceptions of the ends and goods involved, are systematically extended (MacIntyre 1984, 187).

In other words, a practice is a complex activity which involves a group of people and which has certain goals, whose purpose is the promotion of the activity's excellence and maturation.

MacIntyre distinguishes between an external good and an internal good. Now, this is a simple enough distinction: basically, external goods only benefit a single person (or arguably a small group of persons) within the community of a given practice. They are usually material goods or things which one can possess, such as wealth and fame, and whose attainment is not restricted to involvement in a particular practice. Conversely, an internal good is that which benefits both the entire community of practitioners and the practice itself, so that either it upholds the framework governing the practice or lends itself to the betterment and maturation of the practice itself; such goods are only achieved when participating in said practice.

When evaluating healthcare as a practice, there exists an inherent relational and social dimension insofar as healthcare is in and of itself other-oriented as it provides a service that contributes to personal well-being and human flourishing. Healthcare is a practice whose internal goods necessarily and primarily benefit not only the practice itself, or its practitioners, but the community of patients receiving such care, making healthcare a unique practice in that it contains an inherently interpersonal ethical and social dimension.

Writing on virtue and healthcare, the late Catholic bioethicist Edmund Pellegrino proposed a list of virtues of the "Good Physician," which are "Fidelity to trust... Benevolence... Intellectual Honesty... Courage... Compassion... [and] Truthfulness... (Pellegrino 2002)" Commenting on the practical behaviors of the Good Physician, Pellegrino writes:

> It is difficult, for example, to imagine that a physician committed to the virtues I have just outlined would ever consider her relationship with the patient as primarily a commodity transaction, a contract for service, or the mere application of scientific knowledge to a sick organism. A virtue-inspired physician would recognize *pro bono* work as crucial to her stewardship of medical knowledge. Neither could such a physician see herself as an entrepreneur, an investor or as owner of a health care facility like a hospital operated for profit. She would not claim proprietary rights over her knowledge (Pellegrino 2002, 381).

If we come to think of responsibility as a state of being that requires the cultivation of compassion and benevolence and which results in actively engaging in philanthropy and physicianship, then it seems that actively fostering the individual's sense of social responsibility to provide healthcare as a heartfelt response to suffering can play a greater role in medical ethical education. In this way the problems associated with healthcare provision will not simply be reduced to their socioeconomic and political dimensions but can come to bear on our understandings of what it means to be a "Good Physician" in the moral sense. Hence, the issue of philanthropic healthcare can go beyond discussions of how we ought to redistribute material resources to the redistribution of moral responsibility, with both healthcare providers as well as those who educate them in medical ethics partaking in the processes of ensuring healthcare justice.

To this end, I find it practically useful to draw a conceptual moral difference between types of work, especially as applied to the cultivation of virtue and social responsibility in healthcare. Drawing on the Greek notions of *δουλεία*, *ἔργα*, and *διακονία*, I would like to suggest that there exists a distinction between the interrelated concepts of labor, work, and service. We might think of *δουλεία*, or labor, as consisting of those activities that cultivate and sustain biological life and hence contribute to the reproduction and maintenance of human vitality; in this sense authentic labor contributes to the sustainability of natural well-being. Ἔργα, or work, are those activities that build and construct social artifacts, systems, and mechanisms that people use as a means of achieving a state of flourishing and hence, virtuous work becomes a means to achieve higher social goods; from this perspective, virtuous, or just, work will be that which contributes to the cultivation of excellence and human flourishing.

Finally, there is the aforementioned notion of *διακονία*, or service, which comprises actions taken and work and/or labor performed with the explicit goal of fulfilling human needs and thus, entails actively serving the needs of persons and communities. In the context of medicine and healthcare, medical research would fall under the auspices of *έργα* insofar as it entails discovering and developing resources, techniques, systems, and materials with the goal of better enabling human well-being and therefore can be construed as work undertaken and performed with the aim of human flourishing. From this point of view, clinical healthcare can be construed as a form of work or labor aimed at attending to the human need of health and hence, may be thought of as embodying the notion of *διακονία* as healthcare services seek to increase human well-being, restore and maintain human vitality, and contribute to persons' ability to flourish by ensuring that their need for health is fulfilled.

Within this conceptual framework, the idea of philanthropic healthcare will indeed imply that its virtuous individual practitioners engage in pro bono work, as Pellegrino suggests, construed as a form of personal *διακονία* yet must not be restricted to this individual level. We must be cautious not to overlook the communal dimensions of social responsibility and of philanthropy, which are intimately tied to the personal. It is not simply to say that persons always exist in relation, but it is also to suggest that we remain cognizant that philanthropy implies a mutually reciprocal benevolence, which itself contributes to a sense of community among individuals as well as the flourishing of the larger community. In this sense, philanthropic work goes beyond tending to the individual physical needs of persons to addressing their communal and social health needs as well. Therefore, on the social level, philanthropic healthcare involves the creation and establishment of local healthcare centers and possibly nonprofit networks of healthcare providers and clinicians to staff them. This is where the idea of subsidiarity, from Catholic social teaching, can be incorporated into our paradigm of philanthropic healthcare.

Subsidiarity entails empowering local and regional communities capable of carrying out particular tasks and practices themselves to do so by providing them with the means to become self-sufficient as a way of cultivating a deeper sense of community and striving toward genuine communal flourishing. Subsidiarity views the common global good as involving the decentralization of activity and the dispersion of power and action among a variety of local communities. It is in this way that philanthropic healthcare service can foster a sense of solidarity among members of local communities and can possibly encourage the establishment of local charitable healthcare organizations. From the Christian point of view, organizations that send medical teams of volunteers to various regions of the globe to provide healthcare for limited periods of time provide an essentially good and laudable service to the world's sick. However, if philanthropic healthcare is simply reduced to the provision of temporary pro bono work and fails to incorporate the establishment of regional means of continually meeting the healthcare needs of local communities, it will be unable to work toward creating a truly sustainable mechanism for the achievement of health and human flourishing among those being served.

To this end, social action embodying the ethical ideal of *philanthropia* must go beyond immanent issues of illness and disease and address larger systemic issues concerning public health and the environment as well. Christianity shares the ethos found within Article 14, when it calls for all sectors of society to promote health and social development by addressing issues such as the environment and poverty as well as the strictly clinical aspects of healthcare. Between Pope Francis' recent calls for Christians to adopt a more eco-friendly mindset and lifestyle and Patriarch Bartholomew of Constantinople's long-standing advocacy of environmental ethics, for which he has received the moniker "The Green Patriarch," there is an emerging consensus within traditional Christianity that ecological, economic, and health crises are all intertwined and that a fervent alteration of our ways of living in the world as both individuals and as communities is a necessary component of being socially responsible.

In the Christian tradition, the idea of *diakonia* was institutionalized in the diaconate, or the existence of the deacon: an ordained position within the hierarchy of the Church that dates back to the early Christians and which was comprised of both men and women who took vows to serve the community. In the early Church, deacons went out into different communities to administer communion and teach the gospel. By the second century, deacons and deaconesses administered communion, helped spread the message of the gospels though teaching, and tended to the needs of the sick and the poor through nursing and bringing food to the hungry and downtrodden.

The work performed by the Diaconate was so foundational to the Church's presence in the world and was such an integral part of the ecclesial hierarchy that in the second century St. Ignatius of Antioch described deacons and deaconesses as representatives of Jesus Christ on earth. In his Epistles, St. Ignatius discusses the theological nature and significance of the ecclesiastic structures and arrangements of his time and in doing so envisioned the diaconate as a necessary part of Church's mission as well as the Ekklesia's embodiment of Christ. He describes bishops as representatives of God as the Father tending to their spiritual children. He describes the priesthood as representing the apostles spreading the Christian faith and word of God. And, finally, St. Ignatius speaks of the diaconate as representing Christ by performing his ministries of healing, teaching, and selflessly tending to the needs of the sick and the poor.

Although both the Roman Catholic and Eastern Orthodox Churches have maintained the diaconate, the role of the deacon is far less robust than it was in Ignatius' era. While all Christians are called to engage in *diakonia*, and this point cannot be emphasized enough, one way in which diaconical philanthropy can be institutionalized is by reinvigorating the diaconate by encouraging both men and women seeking to serve the Church to pursue ordination as deacons and deaconesses and by expanding the role of these clergy to include, for instance, an educational vocational path, which would go beyond the teaching of the gospels to include teaching moral theology and ecclesial history among other subjects; a vocation of healthcare provider, where deacons and deaconesses trained in medicine could perform medical missions in global society and administer their services to the world's sick; and an ecological vocation, in which deacons could commit themselves to practices and projects addressing environmental issues and problems of public health on both local and global levels in an attempt to strive for systemic change as a way of transfiguring the world.

To conclude, many of the concepts I have discussed, such as philanthropy, charity, hospitality, and service, have lost the richness of their ancient meanings and are often used to refer solely to voluntarily giving financial assistance and being part of the lodging or food industry. When discussing what social responsibility in global bioethics entails, pausing to reflect on the Christian meanings of these concepts can shed light on the ways in which personal and social responsibilities are related and the ways in which they can be employed in the uncoerced pursuit of health as a common good and the revitalization of the civil sphere of global society as the locus of moral action. Speaking as a theologian, John McGuckin has commented that:

> So much of philanthropy in our modern world has returned to the pre-Christian hellenic model. It is motivated by concomitant patters of guilt, accompanied by loathing and neglect for the marginalized… Can there be… a return to a much more encompassing vision of energetic rebuilding of social structures on the basis of remaking a sense of the mystical dignity of the person? The task is deeply theological. It is equally a pressing social demand. The two things are not incompatible…Our task today is one of extensive reconstruction (McGuckin 2010b).

In many ways the suggestions I have made throughout this essay are in a sense a philosophical reflection that responds to this claim and attempts to delineate ways in which Roman Catholics and Eastern Orthodox alike can take up this task of extensively reconstructing their notion of philanthropy as it relates to healthcare on a global level.

In accord with the Christian moral vision yet, going beyond any suggestions that are explicitly directed toward members of the Christian Church, the views I have expressed concerning the relations between our conceptions of virtue and duty, personal and social responsibility, as well as law and ethics are intended to speak to a broader audience comprised of those coming from all religious and philosophical persuasions to reflect upon our immanent need to actively addresses issues of health and healthcare on a global level.

In our pluralistic era in which not all persons are of the Christian faith and hence do not recognize the divinely inspired nature of the decalogue nor the importance of mimicking the divine in their pursuit of the good life, forms of law and sets of rights and duties derived from other sources that are compatible with the ethos delineated in this essay can come to serve as a pragmatic trans-religious and cross-cultural deontology that is capable of serving as a global schoolmaster of virtue. To this end, the principlistic and legal dimensions of the UNESCO Universal Declaration on Bioethics and Human Rights might come to be perceived as a means of fostering the cultivation of public virtues in regard to matters of healthcare and bioethics that all members of the global community can come to share in common regardless of their respective world views or faith traditions.

References

John Paul II. 1984. *Apostolic letter Salvifici Doloris*, No. 28.

MacIntyre, Alasdair. 1984. *After Virtue*. 2nd ed, 186. Notre Dame: University of Notre Dame Press.

McGuckin, John. 2010a. Embodying the new society: The Byzantine Christina instinct of philanthropy. In *Philanthropy and social compassion in the Eastern orthodox tradition*, ed. M. Pereia, 60. New York: Theotokos Press.

———. 2010b. Embodying the new society: The Byzantine Christina instinct of philanthropy. In *Philanthropy and social compassion in the Eastern Orthodox Tradition*, ed. M. Pereia, 68. New York: Theotokos Press.

Patriarch Bartholomew. 1997a. (address to Southern Methodist University) Spirituality and human rights, In *The world yet not of the world*, ed. John Chryssavgis, New York: Fordham University Press.

———. 1997b. (Opening Address, Hailki, Turkey) *On earth as in heaven*, eds. Patriarch Bartholomew, John Chryssavgis, Duke of Edinburgh, 2011, New York: Fordham University Press.

Pellegrino, Edmund. 2002. Professionalism, profession, and the virtues of the good physician. *The Mount Sinai Journal of Medicine* 69 (6): 381.

UNESCO. 2005a. *Article 1, Universal Declaration on Bioethics and Human Rights.*

———. 2005b. *Article 2, Universal Declaration on Bioethics and Human Rights.*

———. 2005c. *Article 14, Universal Declaration on Bioethics and Human Rights.*

Chapter 6
Supererogation and Social Responsibility: A Response to Chris Durante

David Heyd

Abstract In this comment I suggest that the Parable of the Good Samaritan should be understood as distinguishing between three levels of normative guidance: the legal, the moral, and the supererogatory. The distinction between the obligatory and the supererogatory is taken as objective and independent of the kind of motivation or virtue underlying both kinds of actions. However, it is argued that states, as political institutions committed to justice and impartiality, cannot be agents of supererogatory action. Charity, in contrast, may be ascribed to institutions (like charities!) which act by the intention to do more than is strictly required. The final part of the comment examines the Jewish tradition with regard to acts held by the Catholic doctrine as supererogatory.

Keywords Supererogation · The Parable of the Good Samaritan

This is not "a Jewish comment" on the Christian perspective on social responsibility in bioethics but rather an attempt to articulate further one Christian theme or to be more precise – a Roman Catholic one. I will come in the end of the comment to a few points relating to the Jewish interpretation of that theme.

Durante notes in the beginning of his paper the legalistic language of the UN Declaration which takes as its aim to provide guidance and assist in decision-making on bioethical issues. This is only natural, particularly since the Declaration is addressed not only to individuals but also to states and institutions which are typically guided by rules and principles. But, Durante adds, such rule (or right-based) regulation is not sufficient, at least from the Christian point of view. Cultivating a proper ethical culture in medical practice requires also the creation of personal *virtue*, that is to say, a certain motivation and indeed moral attitude rather than sheer action in accordance to some rule. The fundamental ethical point of medical care lies beyond

D. Heyd (✉)
The Hebrew University, Jerusalem, Israel
e-mail: david.heyd@mail.huji.ac.il

J. Tham et al. (eds.), *Religious Perspectives on Social Responsibility in Health*, Advancing Global Bioethics 9, https://doi.org/10.1007/978-3-319-71849-1_6

mere curing or the restoration of some previous state of biological health: it should be seen as providing *service* to suffering people, motivated by love and charity.

Seeing virtue as a necessary complement of justice or seeing an act of justice in the strict sense as requiring a virtuous agent raises the theoretical problem of the relation between action and agent, formal justice and personal attitude, and moral standards and moral motivation. It seems that there is an important strand in Christian theology which takes seriously the distinction between duty and virtue or between moral doing what is (religiously) required and what goes beyond it, between moral and saintly acts. The textual starting point for the discussion of this distinction and for the doctrine of supererogation which ensued is the Parable of the Good Samaritan in the New Testament, which is mentioned by Durante but interpreted in a different way from that which I propose.[1]

I want to focus on the last sentence of the parable, in which after bringing the injured man to the inn and taking care of his needs, the Samaritan tells the innkeeper: "Look after him, and on my way back I will make good any extra expenses you have." The Latin *quodcumque supererogaveris* means literally "whatever you pay extra" or "more than is due." This little phrase has made a glorious theological career when already the Church Fathers have introduced the technical concept of *supererogation* as referring to actions that go beyond the call of duty. Thus, St. Ambrose (n.d.) and St. Augustine (n.d., p. 157) draw the distinction between *praecepta* and *consilia*, commandments and recommendations, and connect this distinction to other stories in the New Testament in which Jesus is said to have implied that distinction. Thus, poverty, chastity, and absolute obedience (i.e., monastic life) became known as the three evangelical counsels. Those who decently earn what they possess have satisfied the commandment, but those who give all they have to the poor do more than they have to do. They act supererogatorily. The same applies to chastity and obedience. People who get married and beget children are doing their religious duty, but those who choose a life of chastity do more than is strictly required. Worshipping God according to the rules of the Church makes one a good Christian, but committing oneself to total obedience in which one's whole life is devoted to God is superior but morally optional.

Thomas Aquinas gave the idea of supererogation a full systematic theory and expanded its scope beyond the three evangelical counsels of poverty, chastity, and obedience. Works of supererogation are those which it is virtuous to do but not absolutely required. They express the difference between the Old Law (of duty) and the New Law (of love). They introduce the element of freedom (*licentia*): acts of particular generosity are voluntary, a matter of individual discretion. They aim at perfection which can be achieved only by transcending the requirements of the law. And they deserve extra merit which is most typically manifested in the lives of *saints* whose life is consistently guided by supererogatory behavior.[2] Thus, during the thirteenth century, the idea of supererogation became institutionalized: the extra

[1] Luke 10:25–37.

[2] St. Thomas Aquinas' views on supererogation can be found mainly in *Summa Theologica*, the first part of the second part, questions 100, 106, 107, 108, and 184 as well as in *Summa contra Gentiles*, book III, part II, chapter 130 (1928, 1947–8).

merit of saints, of actions which are "extra" from the religious point of view, is kept in the so-called spiritual treasury of the Church and dispensed by the Church for compensating those human beings for having failed to reach the necessary required level, thus enabling them to attain eternal life.

But this institutionalization, with the system of *indulgences*, which since the Crusades has become commercialized and had a corrupting effect, led to the downfall of the whole idea of supererogation during the Reformation. Luther, Calvin, and the Anglican Church all abhorred it. Supererogation seems to have lost its central role in the Church teachings in the last few centuries.

But strangely enough, the idea has been revived by secular philosophy. In 1958 the Oxford philosopher James Urmson published an article "Saints and Heroes" and used the concept of supererogation as a fourth category, beyond that of the obligatory, the permitted, and the forbidden: acts that are morally good but not obligatory (Urmson 1958). Since then, supererogation has become a never-ending subject of sophisticated debate in moral theory. Like Protestant theology, there are opponents to the very idea of morally valuable actions which are not required. But I believe that the concept of supererogation is important and useful both analytically and morally.

Let's go back to the old road from Jerusalem to Jericho. I understand the Parable of the Good Samaritan as describing three alternative ways of passersby to respond when crossing a wounded man lying in their way. They can ignore him, as the Priest and the Levite did. They can help him (put a bandage on his wounds and carry him to the nearest inn). And they can, in addition to such help, come on the next day and visit him, paying for all the costs of his treatment and accommodation. The Priest and the Levite behave within the requirements of the strict *law*. Except for few legal systems today, most countries do not require by law to help a person in need or distress.[3] The person lending help to the victim and leading him to a safe place acts in the way that a *moral* person is expected to act. They do more than is legally required of them, out of a sense of duty or as a matter of conscience. But the Good Samaritan "supererogaverit," that is, pays out more than could be ever expected even of a morally decent person. He goes out of his way, literally and metaphorically, in his love for the neighbor, a total stranger.[4]

This brings us to the specific subject of moral responsibility. The provision of health has become a collective and highly regulated practice. It is no more the individual doctor committed by the Hippocratic Oath to give equal, and sometimes free, services to individual patients. The state is typically an entity which can only act by law and is bound by special responsibility to its citizens. Hence I find Durante's attempt to combine the UN Declaration with the cultivation of virtue problematic. States and other political or social institutions (including hospitals) do not act with compassion or out of personal care. They lack character and emotion. They cannot – and maybe ought

[3] There are only a few countries in which lending assistance to people in distress is legally required. Good Samaritan laws (as they are called) are morally controversial and difficult to enforce (as such a law in Israel has proven to be). For an old but important collection of articles on the subject, see Ratcliffe (1966).

[4] For a fully fledged theory of supererogation, see my *Supererogation* (Cambridge: Cambridge University Press, 1982).

not – act out of empathy or mercy but rather operate on the basis of standards of action, institutional aims, and respect for the rights of others. States are not moral agents acting with good intention or out of conscience or altruism.

And once we consider helping other countries in need (in contrast to one's own citizens), the operation of states cannot even be legally regulated because even if we can imagine international law regarding war or respect for sovereignty, there is no parallel law regarding the global distribution of medical resources. Allocating state money for international aid is at most a moral responsibility, usually motivated by the concern of individual citizens in a well-off country for the afflicted citizens of a poor country. These citizens may decide by a democratic vote to allocate a particular percentage of their GDP for foreign aid. But with so many citizens lacking access to decent medical treatment in their own country, it would be contrary to the idea of state responsibility not to give them priority before assisting foreign countries.

But although I find it odd to ascribe supererogatory action to states, it is definitely the case that *charity* institutions act by definition in a supererogatory way. They do what states fail to do. They work on the basis of voluntary donations which are not strictly required, even on the moral level. And more typically, they work through individuals who do much more than could be morally expected of a morally decent individual. They often risk their lives, give up the pleasures of comfortable existence, and devote years of their life to faraway people. Mother Theresa is a proverbial example. Doctors Without Borders is another institution which illustrates the idea of going beyond the call of duty. And as Durante mentions, the Church is a large nongovernmental institution providing health services in the whole world. No wonder that the same word "charity" refers to both institutions and an individual virtue which lies beyond justice and rights. My main claim in support of supererogation is that exactly because our conceptions of justice in the distribution of health services on the international level are so controversial and problematic, supererogatory behavior becomes both necessary and noble.

Agents of supererogatory acts often report that they only did what they felt they "had to do." Although they believe that their behavior was in some sense "required" or called for, they do not impose that requirement on others. This is exactly the sense in which their action is supererogatory rather than obligatory despite their subjective perception of it. The line between the obligatory and the supererogatory is objective; that between what *I* feel I ought to do and what others ought to do is merely subjective. The fact that supererogatory action manifests the pursuit of perfection or is a response to some calling or vocation does not mean that it is not objectively beyond the call of duty.

Finally, a word about the Jewish tradition on this subject should be added. Indeed, like Protestantism, Jewish religious thought does not have a parallel concept of evangelical counsels. Religious morality is based strictly on justice and hence on duties and commandments. However, these include the commandment of "love thy neighbour as thyself" which goes back to the Torah and hence much earlier than Christianity. On the other hand, the Christian interpretation of that commandment is open, has no limits, and hence lends itself to supererogatory generosity, like that of the Good Samaritan. It also universalizes the very concept of "neighbor" to any human being as such. But despite the image of the strictness of adherence to the law, there is the concept of *ex gratia* ("lifnim mi-shurat ha-din"), namely, doing something although not required by law or even bending the law when there are good moral reasons to do so. There is also the idea of forgiveness and mercy which logically lies beyond the rules of justice and can be understood as having a supererogatory nature. Compassion and charity are considered highly virtuous, and although there are no saints in the Jewish tradition, there are "righteous" people who in some way do more than is required. There is also the character of the *hassid*, the pious man who is engaged in prayer more than is required of others.[5] However, there is no ideal of supererogatory perfection such as monastic life, chastity, or extreme charity which leaves a person with no property. Social responsibility of both individuals and the community – both in general and in the provision of medical help – is articulated in terms of binding rules. Nevertheless there is a well-known Talmudic saying: "Jerusalem was destroyed [two thousand years ago] only because they based their judgments [strictly] upon Biblical law, and did not go beyond the requirements of the law."[6] This can be interpreted as recognition of some form of supererogation but can equally be understood as the requirement of judges to make their decision according to the spirit of the law rather than its literal meaning.

The unprecedented ideal of the universal community of all human beings is an original contribution of Christianity. It is no coincidence that the idea of human rights developed in the framework of Christian culture (although only a millennium and a half after the New Testament). However, the Jewish notion of moral responsibility in areas such as the provision of health care is typically community-based, but not in the universal sense. A community is defined as essentially particularistic. Responsibility, like solidarity, applies within a given group, formed by some historical, religious, or political circumstances or choices. Consequently members of the group have priority in the way a responsible community or state provides them with services – including medical services. As the Talmudic dictum goes, "the poor of your community take precedence over those of other communities."[7] The UN Declaration can be understood as imposing duties on member states to care for their respective citizens and provide them with health services in a fair and effective manner and even as imposing duties on richer states to lend assistance to poorer states in advancing health care. But this does not mean that duties of individuals, institutions, and states are *universal*, including in the case of the provision of health and medical care. Beyond a certain degree, such assistance is supererogatory.

References

Ambrose, St. n.d. *Concerning Women*, Chapter 12. New York: Fathers of the Church.
Augustine. n.d. *Letters*. New York: Fathers of the Church.
Ratcliffe, James M., ed. 1966. *The Good Samaritan and the law*. New York: Anchor Books.
Urmson, J.O. 1958. Saints and heroes. In *Essays in moral philosophy*, ed. A.I. Melden. Seattle: University of Washington Press.

[5] Babylonian Talmud, Tractate *Berachot*, p. 32b.
[6] Babylonian Talmud, Tractate *Baba Metzia*, p. 30b.
[7] Babylonian Talmud, Tractate *Baba Metzia*, p. 71a.

Chapter 7
Virtuosity Beyond the Call of Duty: A Reply to David Heyd

Chris Durante

Abstract In response to David Heyd's article concerning supererogatory actions, I will attempt to make sense of the notion that there are degrees of goodness by proposing the foundations of a theory of the good which incorporates both duty-based and virtue-based understandings of ethics. Drawing from moral philosophy as well as Christian moral theology, I will argue that a deontological framework establishes the criteria for moral decency while a virtue-based framework establishes the criteria for moral goodness. Hence, while supererogation does indeed become redundant in a strictly virtue-based paradigm of ethics, insofar as supererogatory acts would be virtuous acts on such an account, a virtue ethics concerned primarily with the good can still make room for a pragmatically inspired deontological system to be enacted as a means of managing the behaviors of all of those persons who are less than virtuous. In Christian ethics, this implies that the divine moral laws are not "right" because they are divinely commanded, as some theorists may suggest, but rather because they are the best instruments that persons can use to become good or cultivate goodness in their lifetime; they are the best tools we have on the path toward becoming Godlike or Christlike. On this interpretation, Christian ethics is not a form of divine command theory but is rather a virtue ethics that also encompasses a pragmatically and pedagogically valuable deontology within its moral purview.

Keywords Supererogatory · Duty · Virtue · Moral decency · Moral law

In response to David Heyd's ideas concerning supererogatory actions, I fully understand that in a deontological-based framework, duty—and thereby its fulfillment—tends to be the measure by which a person is judged to be good. If morality is primarily about duties and interpersonal obligations, then one whom correctly understands and upholds such duties and performs her obligatory actions to other members of society may indeed be considered a "good person." On this model a

C. Durante (✉)
Department of Theology, Saint Peter's University, Jersey City, NJ, USA
e-mail: cdurante@saintpeters.edu

J. Tham et al. (eds.), *Religious Perspectives on Social Responsibility in Health*, Advancing Global Bioethics 9, https://doi.org/10.1007/978-3-319-71849-1_7

"virtuous person," or one who is characteristically "morally excellent" (as "virtue" and *arete* imply), may be considered one who has gone beyond the call of duty and, hence, surpassed the ordinary marker of standard goodness. This then implies that there exist different levels, or degrees, of goodness itself and presents us with the task of figuring out how we should go about determining the nature of this "hierarchy of goodness," so to speak. While I am in no way attempting to refute the notion that there exist many dimensions and degrees of goodness, I am reticent to acquiesce to the idea that we are even remotely capable of understanding and establishing such a hierarchy of goodness by simply assessing the nature of actions and obligations.

A holisitic philosophy of the good will encompass both duty-based and virtue-oriented modes of ethical analysis within its general paradigm. Strictly deontological paradigms of moral reasoning tend to allow for "rule followers" to be considered "good" regardless of their sentiments, dispositions, and affective motives for action. A related problem is that virtue theory has come to be considered a stand-alone school of moral philosophy, either construed as somehow diametrically opposed to deontological ethics or so narrowly construed that discussions of virtue have become dissociated from the original Aristotelian project of developing a philosophy of the good, within which the concept of virtue is operative. If we begin with a consideration of goodness and take the term virtue in its most authentic sense to imply "excellence," then that which is to be deemed "virtuous" must reside at the top of our hierarchy of goodness. Consequently, if virtue theory entails the discussion of intentions, dispositions, and habits in addition to actions and behaviors, then it necessarily entails discussions of moral psychology and cannot be divorced from an agent, or person, centered approach to moral analysis. Yet, if persons and their lived psychosocial realities are indeed to be the basis of our ethical analysis, we must remain cognizant of the fact that in lived social reality, persons' modes of moral reasoning and life experiences are comprised of a moral bricolage of duties, consequences, and virtues. The moral life of real persons cannot be truly understood if we attempt to analyze these notions as if each exists in an abstract theoretical vacuum. For proper philosophical analysis, our discussions and debates cannot take place as if this were the case.

Additionally, in order to achieve the most accurate understanding of moral life, our ethical analysis of social responsibility and moral judgements of persons must not be overly simplistic and become mired in binary modes of thought. This is arguably the contemporary situation that we must seek to overcome. While on the one hand ethics has grappled with the existence of multiple traditions of moral reasoning and has indeed begun to incorporate insights from psychology and the social sciences, neuroscience, and the cognitive sciences as well as other fields in recent years, the majority of academic moral thinking remains mired in the simplistic binary that has been constructed between good and evil, or moral and immoral, as exclusive categories of moral being at the expense of a more nuanced and dimensional approach to characterizing human behavior and agency. Although it seems perfectly reasonable to describe actions as being good, bad, or neutral—as Catholic moral theology has traditionally done in the case of describing and categorizing actions—the category of "neutrality" seems to be rather out of place if used to describe the moral character of human persons. We need to broaden our modes of discourse and the categories of morality that we discuss. One problem is that we rarely speak of decency anymore. We spend a considerable amount of time attempting to describe "goodness" or "evil" or attempting to define the "right" or the "wrong." Yet, the question I pose is what does it mean to be a "decent person"? Decency as a moral category exists on the spectrum somewhere between goodness and badness, rightness and wrongness, or righteousness and depravity. Ultimately, by re-focusing our attention on decency, and its relation to virtue, it is possible to reframe the relation between deontological morality and virtue ethics and come to view these two modes of moral reasoning as compatible.

If we begin with a philosophy of the good consisting of a virtue-based method of assessing person's character traits, and the virtuous person is construed as the ideal of what it means to be the "good person" (that all are expected to strive to be like), we are presented with an ethics in which "supererogation" is expected, so to speak, and hence supererogatory acts would cease to be "supererogatory" in this sense. Yet, as I have been arguing, our task is not to pit deontology against virtue in a zero-sum intellectual match as if moral life consists in either one or the other; the robustness of moral life is comprised of both. Thus, what I have in mind here is a conception of goodness, which, while still admitting of degrees, is grounded in virtue yet which does not neglect deontology. In this schema duties would constitute a minimum standard of moral activity required of each person, while supererogatory acts would be encouraged for every person in conjunction with their pursuit of becoming virtuous. What I wonder is whether or not there is a way to retain Heyd's concerns regarding the degrees of goodness, so to speak (which the concept of supererogation implies), while incorporating insights from both deontological and virtue-based ethical paradigms in our understanding of Christian ethics.

I think this has a lot to do with the ways in which we use and conceptualize the word "decency" or "decent person." In a system in which both virtue and duty are operative, in which virtue is the gold standard of goodness, decency may not be all it is cracked down to be. What I mean here is that she who upholds her minimum moral duties and obligations would constitute an ethically decent person. Hence, simply abiding by moral laws/upholding duties for whatever reason and without necessarily having cultivated particular dispositions such as compassion, courage, generosity, etc. would be the mark of a morally decent human being yet not the behavior of a morally good person. From the point of view of a modest and pragmatically efficacious Christian social ethic, the alleged tension between deontology and virtue-based approaches to ethics is relieved if we incorporate the category of moral decency into a multi-tiered system of ethics that relegates the cultivation of decency to the arena of social deontology and which views deontology as a means for cultivating virtue. Presupposing there exist certain moral laws and duties, such as prohibiting murder, we might imagine an individual who refrains from murdering yet does so out of fear of punishment or cowardice, for instance, and not because she has cultivated any sort of disposition that would make her experience a negative sentiment when contemplating murdering or a sense of empathy with either the humanity or circumstances of the innocent person she is thinking of killing. Yet, can we truly call a person who refrains from murder because of some fear of punishment or retribution or cowardly disposition a good person? It seems as though such a person would be better described as a decent person, but not as someone who is worthy of emulation, as is the "good person" in a virtue-oriented ethical paradigm. In sum, what I am suggesting is that moral decency entails upholding minimal duties/obligations while moral goodness entails acting virtuously, which necessarily entails the intentions and sentiments motivating one's actions and not simply an act as a behavioral performance.

In the Greek Orthodox Christian tradition, there is a saying that "Law is the schoolmaster of virtue." This statement implies that the decalogue, or the divinely issued ten commandments, is indeed an important feature of the ethical life; however, to simply uphold the commandments and abide by the laws without simultaneously experiencing a heartfelt transformation of one's desires and dispositions so that one sincerely wishes to bring about a world in which love of the divine and neighbor are a reality will not suffice if the ultimate telos, or goal, is the transfiguration of both persons and the world in the image of a loving and benevolent God. Hence, the deontological dimension of Christian morality serves the purpose of directing persons toward morally correct and righteous behaviors that will in turn assist in fostering dispositions to be the type of character who holds a genuine concern for the well-being of others and the world and an authentic love of the divine. From this perspective, the law of God is not strictly a contract but an instrument of personal moral cultivation and self-development. Consequently, if each individual is not "contracted" to assist another but is rather called to cultivate a sincere love of neighbor, morality takes on a very personal dimension in that it requires self-development.

Supererogation only becomes a "problem" for virtue-based modes of moral reasoning if virtue is believed to be an exhaustive or all-encompassing category of moral action (or the only mode of moral assessment by which we are able to evaluate all behavior). An ethicist concerned with cultivating the virtues can maintain that it is only through a virtue-based paradigm that we can establish the good while simultaneously maintaining that some form of deontology is necessary in social life as a means of coping with persons who are not quite "virtuous" per se yet who we would not consider to be "immoral" in any salient sense of the term.[1] Ultimately, while it is only the "virtuous person" who is to be deemed "good," in the most robust sense of the term, goodness is not an exhaustive category of moral being. A decent person is neither "good" per se nor "evil" yet may be said to customarily act rightly.

Consequently, *a deontological framework establishes the criteria for moral decency, while a virtue-based framework establishes the criteria for moral goodness.* Hence, while supererogation does indeed become redundant in a strictly virtue-based paradigm of ethics, insofar as supererogatory acts are virtuous acts, a

[1] We might all have our moral shortcomings; however, this is not the type of "immorality" I have in mind here.

virtue ethics concerned primarily with the good can still make room for a pragmatically inspired deontological system to be enacted as a means of managing the behaviors of all of those persons who are less than virtuous. Our tasks as ethicists, on an account of social ethics in which deontology is considered the "schoolmaster of virtue," do not entail deriving duties and obligations from character traits and dispositions but rather entail creating a set of social duties and obligations that help cultivate the habits required to become virtuous.

If deontology is the schoolmaster of virtue, then it would seem that we need a solid understanding of moral psychology not only to establish the set of virtues we ought to endorse, as Anscombe has suggested in the past, but also in order to establish a deontological code of decency that will simultaneously be the most useful code to regulate immoral conduct as well as one that will be useful in cultivating virtuosity in those who abide by it. In this way, the good will only come before the right conceptually but also after the right pragmatically; on this account the good is both the A and Ω of the right. In Christian ethics this implies that the divine moral laws are not "right" because they are divinely commanded, as some theorists may suggest, but rather because they are the best instruments that persons can use to become good or cultivate goodness in their lifetime; they are the best tools we have on the path toward becoming Godlike or Christlike. On this interpretation, Christian ethics is not a form of divine command theory but is rather a virtue ethics that also encompasses a pragmatically and pedagogically valuable deontology within its moral purview.

When discussing the issue of humanly established legal and ethical codes, regardless of how one arrives at such a deontological system of moral laws, ethical norms and social duties, such a system of precepts and codes functions primarily as a means of managing social behavior yet, cannot demand or require persons to adopt particular motives or intentional states; it is a practical impossibility for legal and ethical codes to ensure the alteration of personal dispositions or psychological states. In any deontological framework, it is necessary for an agent to adhere to the norms delineated if she/he is to be considered one who acts rightly.

However, the simple observable fact that an agent consistently adheres to laws and precepts does not necessarily imply that the agent is doing so for the same reasons that said norms we established in the first place; the agent can act in accord with a norm or precept while still holding bad motives for doing so. For example, we may justify and legitimate a law prohibiting murder because of a belief in the common good and the immorality of taking innocent human life, yet a social agent may simply refrain from murder—thereby acting in accordance with the law—simply by sheer fact that they do not wish to get caught and punished or be perceived as "bad" by others, etc. All that a system of laws or code of precepts can reasonably be expected to achieve is a decent social order in which all law abiders are characteristically decent as a minimum standard. Ultimately, deontology sets the standards for moral decency, while virtue sets the standards for moral excellence. It would be unreasonable to expect all persons to be able to achieve moral excellence yet is pragmatically feasible and reasonable to expect all to at least be morally decent.

From the perspective of traditional Christianity, the morally excellent person is one who has the capacity to sympathize with the plight, pain, circumstances, and dilemmas faced by both she/he who adheres to moral laws and she/he who breaches them. The virtuous person is one who is successfully striving to constantly embody the ideals of compassion, benevolence, and philanthropy toward living beings and hence, will go out of his/her way to assist, comfort, and be with others in the time of need. Being truly sympathetic to the suffering of another living being means one's reasons for action will not be motivated by a sense of duty but rather by a heartfelt response to another's situations and circumstances. Moral laws and obligations may serve as an action guide for one on the path toward virtue, yet the morality of being and being moral in the most personal sense entail striving to be like—and becoming—a moral exemplar and thus, require a set of dispositions that the activity of upholding or abiding by rules is unable to achieve on its own. The proclivity to go beyond the demands of moral duty is characteristic of the genuinely virtuous person—the one who sincerely responds to the needs of others with a compassionate presence and whose desire to help them is authentic. If philanthropic sympathy is what drives one to go above and beyond the call of duty, then it is philanthropy that becomes the litmus test for virtuosity and, hence, moral goodness. And, this is the moral message that the parable of the "Good Samaritan" embodies and lesson that it holds for pursuing a Christian lifestyle.

Part III
Confucianism

Chapter 8
Global Standards of Minimum Decency in Health: Social Responsibility and Health Care from a Confucian Perspective

Ruiping Fan

Abstract This chapter argues for a Confucian perspective on social responsibility for health by commenting on Article 14 of the *UNESCO Universal Declaration on Bioethics and Human Rights*. It first indicates that Confucian tradition is supportive of the general ideal that it is a central purpose of governments to promote health and social development for their people. Then it points out that Confucianism cannot accept a radical egalitarian view of social responsibility for global health care. Rather Confucianism would adhere to a view-based concern to promote an adequate minimum of health care for all the people in the world. The chapter concludes by offering two specific suggestions on how to improve health care globally.

Keywords Confucianism · Radical egalitarianism for health care · Social responsibility for health

8.1 Introduction[1]

This chapter offers a Confucian perspective on social responsibility for health by commenting on the statements provided in the Article 14 of the *UNESCO Universal Declaration on Bioethics and Human Rights* (UNESCO 2005). Within the Article, 14.1 states that "the promotion of health and social development for their people is a central purpose of governments that all sectors of society share." Moreover, 14.2 asserts that "progress in science and technology should advance: (a) access to

[1] The first draft of this paper was presented at the Fourth International Bioethics, Multiculturalism and Religion Workshop in Mexico City on November 11, 2014. I wish to thank two formal respondents, Roland Chia and Peter Au Yeung, for their written comments on my paper. I am also grateful to the other participants in the workshop, especially Michael Baggot, Joseph Tham, and Alberto García, for their useful suggestions for revision.

R. Fan (✉)
Department of Public Policy, City University of Hong Kong, Kowloon Tong, Hong Kong
e-mail: safan@cityu.edu.hk

J. Tham et al. (eds.), *Religious Perspectives on Social Responsibility in Health*, Advancing Global Bioethics 9, https://doi.org/10.1007/978-3-319-71849-1_8

quality health care and essential medicines, especially for the health of women and children, because health is essential to life itself and must be considered to be a social and human good." Finally, 14.2 also clearly sets forth a universal right to health: "the enjoyment of the highest attainable standard of health is one of the fundamental rights of every human being without distinction of race, religion, political belief, economic or social condition."

In the second section, I indicate that Confucian tradition is supportive of the view as stated in 14.1. Then I conduct a critical analysis of the ideas offered in 14.2 in the third section. I first point out that for "a right to health" as covered in 14.2, there can be two different interpretations: one radically egalitarian and other decent minimum. Each of them is apparently possible based on the literary expression that "the enjoyment of the highest attainable standard of health is one of the fundamental rights of every human being" stipulated in 14.2. The fourth section shows that Confucianism has strong moral considerations to reject the radical egalitarian interpretation, and I summarize such considerations in the section. Finally, I offer two specific suggestions about how to implement a global right to a decent minimum standard of health in the concluding section.

8.2 Government Purpose

Confucianism would support the view as expressed in 14.1. I will offer my Confucian account for this view by drawing on classical Confucian moral and intellectual resources, especially those provided in the *Mencius* by a most influential classical Confucian writer, Mencius (372–289 BCE).[2] Basically, from Mencius' view, it is from the Mandate/Dao of Heaven (*tianming*, *tiandao*) that there should be benevolent governments (*renzheng*) to rule human society. Such governments exist for the benefit of people and are justified by their ability to protect and promote the well-being of people. This view of political purpose has been termed the Confucian "service conception" of governments: Heaven did not create the people for the sake of government; Heaven established government for the sake of the people (cf., Chan 2014, p.30).[3] Accordingly, given that in the Confucian view the purpose of governments is nothing but serving and promoting the interests of people, Confucianism would certainly support that "the promotion of health and social development for their people is a central purpose of governments that all sectors of society share" as stated in 14.1.

[2] The *Mencius* is selected for the sake of offering this account not only because the *Mencius* has been taken as one of the four basic Confucian books ever since the Song dynasty of China in the eleventh century but also because it provides a detailed account of the central Confucian principles directing the policy formulation and operation of Confucian benevolent governance, as I will show in the text. Presumably, at least regarding moral issues around health and health-relevant services, this account of Confucian benevolent governance should essentially be uncontroversial among various Confucian classics, schools, and figures.

[3] This service conception "is clearly in opposition to the ownership interpretation of *tianming*, namely, that *tianming* grants the ruler an ownership right to the land and people" (Chan 2014, p.31).

Moreover, in order for governments to promote their people's health and development properly and effectively, Mencius upholds a few fundamental Confucian metaphysical and moral principles to direct the policy formulation and administration of governments. First, governments must recognize and appreciate the fundamental importance of the family for the well-being of the individual. "There is a common expression," Mencius points out, "'the world, the state, the family.' The world has its basis in the state, and the state in the family" (Mencius 4A.5).[4] As humans are naturally born in, grown up, and cared for in families, Mencius, following Confucius (551–479 BCE), endorses a foundationally Confucian family-based and family-oriented virtue (*de*), *ren*. *Ren* calls for individuals to practice universal but non-egalitarian love through learning and exercising familial and social rituals (*li*): while one is required to love every human being through appropriate ritual performance, one has metaphysically grounded and ritually distinguished moral obligations to look after one's family members first and more than others (see next section for relevant issues about health-related concerns). Accordingly, Confucianism endorses differentiated and graded love (*chadengzhiai*) rather than egalitarian love (*pingdezhiai*). In short, from the perspective of this fundamental Confucian virtue of *ren*, it is impossible for governments to promote their people's health and well-being without protecting the integrity, stability, and prosperity of the families of their people.[5]

Mencius recognizes that the integrity, stability, and prosperity of families will inevitably be destroyed if families do not have sufficient material means to take care of their members:

> The people...will not have constant hearts if they are without constant [material] means. Lacking constant hearts, they will go astray and fall into excesses, stopping at nothing. To punish them after they have fallen foul of the law is to set a trap for the people. How can a benevolent man in authority allow himself to set a trap for the people? Hence when determining what means of support the people should have, a clear-sighted ruler ensures that these are sufficient, on the one hand, for the care of parents, and, on the other, for the support of wife and children, so that the people always have sufficient food in good years and escape starvation in bad; only then does he drive them towards goodness; in this way the people find it easy to follow him (Mencius 1A.7).

At Mencius' time, the lords of the states levied heavy taxations and corvees on their people for enhancing their own income to pamper themselves in luxury lives and strengthening their military forces to conquer other states. Mencius vociferously condemned such policies and took pains in attempting to persuade the lords to change their policies and turn to benevolent governance. He argues that benevolent governments must make their taxes and levies light. In particular, he followed Confucius again to contend that governments ought to impose only "a tax of one in ten and to abolish custom and market duties" (Mencius 3B.8). Only in this way, he contends, can families be left free to pursue their wealth through productive labor in society and voluntary exchanges in the market. Then the people in local communities

---

[4] All my citations of the *Mencius* in this essay are adapted from Mencius 2003, translated by D.C.Lau.

[5] See Fan 2010 for more detailed argument for this line of consideration.

can be taught and cultivated to "befriend one another both at home and abroad, help each other to keep watch, and succor each other in illness" (Mencius 3A.3). Consequently, "they will live in love and harmony" under benevolent governance (Mencius 3A.3).

This Menciusian quasi-libertarian view remains heuristic to contemporary governance. Although no powerful government today would confess that its tax policy is aimed at maintaining the enjoyments of its own government officials or enhancing the force of its military power to repress other countries, reality frequently tells a different story. A fashionable modern ideology emphasizes that government offer welfare to its people, especially to its most unfortunate members as a priority. But a crucial issue lies in who should be taken as the most unfortunate members in society. Fitting in the cultural features of Confucian familism, Mencius answers this way: "Old men without wives, old women without husbands, old people without children, young children without fathers – these four types of people are the most destitute and have no one to turn to for help. Whenever King Wen put benevolent measures into effect, he always gave them first consideration" (Mencius 1B.5).[6] This is to say those individuals without complete families are the most unfortunate members of society to whom government ought to provide assistance. This is to say, differing from the pure libertarian point of view that government should provide no welfare to anyone from public funds, Mencius would support establishing a safety net to ensure a basic level of health and welfare for those who do not have complete families in society. Now how egalitarian should such a safety net be to everyone in society? To this issue I now turn by discussing 14.2.

8.3 Two Interpretations of a "Right to Health"

Given the Confucian ideas explained above, it would be proper for Confucians to endorse that "progress in science and technology should advance: (a) access to quality health care and essential medicines, especially for the health of women and children" as stated in 14.2. This is not only because Confucianism is in line with the judgment that "health is essential to life itself and must be considered to be a social and human good" but is also because the purposes of developing science and technology under a Confucian benevolent government would certainly include that of advancing its people's "access to quality health care and essential medicines." Finally, given Confucian family-based and family-oriented ethics, putting emphasis on "the health of women and children" would largely be reasonable, because women and children are generally the weaker members both inside and outside of families.

However, having said that, it is not clear if Confucians can readily accept the view of social justice as well as the stated right to health as contained in 14.2. This depends on how egalitarian the whole statements of 14.2 actually are. No doubt, these statements adopt a global, rather than statist, theory of justice regarding human

[6] King Wen was a sage king living in the early Zhou Dynasty (c. 1000 BCE), whom both Confucius and Mencius highly admired.

health – that is, a human right to health should be applied globally, not merely locally. Moreover, the entire Declaration recognizes that human well-being is comprehensive in nature. While health is by itself an essential dimension of human well-being, many other dimensions, such as nutrition and education, can significantly affect health. Thus, simply emphasizing health care is inadequate for protecting individual health. Accordingly, 14.2 properly focus on *a right to health*, rather than *a right to health care*. A right to health would imply a broader range of moral obligations to protect health than a mere right to health care. Both this point of stressing a right to health, as well as the emphasis placed on global rather than local justice, are considerably legitimate from a Confucian perspective.

The problem we confront is what kind of a right to health is precisely stipulated in the statement that "the enjoyment of the highest attainable standard of health is one of the fundamental rights of every human being" as stated in 14.2. This statement is open to two different interpretations, and each of them is apparently possible based on the literary expression of the statement. The crucial issue consists in how "the highest attainable standard of health" should be understood. It could, first, be understood as "a decent minimum standard of health" that is attainable to all human beings in the world. Under this interpretation, the right to health endorsed in 14.2 is that everyone in the world, regardless of one's national citizenship, has a fundamental right to enjoying the decent minimum standard of health. Evidently, this right is plausible and justifiable. On the one hand, as the world has been globalized in various respects, there should be certain moral obligations that extend across national borders to address international health problems. On the other hand, however, such obligations do not have to require that all human beings in all national states have *equal* enjoyment of or access to all services and goods contributive or beneficial to health, as long as they are ever available to any human being in one national state. For example, under this interpretation, we do not have to require that all frontier high-tech medical facilities available to any Americans must also be provided to all Africans. Insofar as a decent minimum standard of health can be reasonably and legitimately established (by an international institution such as WHO), and certain relevant minimum goods and services beneficial to health (such as public health measures, sanitation, clean drinking water, and the like) can be ensured to all humans over the world, this right is satisfied, even if inequalities in health and health-relevant services exist across national states.

However, "the highest attainable standard of health" could also be interpreted to mean "the equally-applied highest possible standard of health," rather than "a decent minimum standard of health." Of course everyone understands that human resources are limited and that we cannot distribute all resources to health-relevant areas only. Even if we adopt a highest possible standard of health, it is impossible to meet all health needs for all human beings all over the world. These points are uncontroversial. Nevertheless, the equality requirement is different. It does not require meeting all the health needs of all people, but only requires meeting health needs *equally* for all people. Essentially, there is no conceptual contradiction to require that all available resources useful to health, no matter how much their total amount is (which is always determined by multiple factors), should always be distributed equally according to health needs all over the world, no matter whose needs they are – this is certainly a theoretically possible way to pursue everyone's enjoyment of health under the highest possible standard of health. Indeed, one can readily argue that for the concern of health, the criterion of national citizenship is as morally arbitrary as the criterion of race, class, or gender. Accordingly, based on this interpretation, if any frontier high-tech medical facilities available to Americans are genuinely beneficial to health and thus should be incorporated into the scope of the highest possible standard of health, then they should also be provided to all Africans as well, regardless of their expenses. As a result, under this interpretation, 14.2 stipulates that everyone in the world has a fundamental right to enjoying the equally applied highest possible standard of health, rejecting inequalities in health and health-relevant services existing across national states.

Many would argue that this idea of a right to the equal standard of health all over the world is practically unfeasible. "Unless the world's economic systems are radically revised, this conception of a right is utopian" (Beauchamp and Childress 2013, p.272). However, the feasibility problem aside, moral reasons for a right to health should first be considered. Do we have convincing moral reasons to support this radical global egalitarian idea? If we do, we should manage to revise the world's economic systems according to the idea and strive to apply a highest attainable standard of health equally to all the people in the world. No matter how hard this work is, we ought to move in that direction because it is the morally right thing to do. On the other hand, however, if we have convincing moral reasons to object to this idea of a right, then we should reject this radical egalitarian interpretation and should instead advocate the "decent minimum standard" interpretation as mentioned above. If so, we should instead pursue a worldly health delivery system that allows the existence of inequalities across national states.

8.4 Confucian Justification

I think Confucianism offers strong moral considerations to reject the radical egalitarian interpretation. As shown in Section II, the central principle of the Confucian virtue of *ren* is differentiated and graded love, rather than radical egalitarian love. I think this Confucian virtue of love can be expressed in the three following Confucian convictions on different moral obligations: in general situations,

1. One has more moral obligation to take care of one's family members (such as one's spouse, parents, and children) than others in one's local or religious community (such as neighbors, friends, and acquaintances).
2. One has more moral obligation to take care of those in one's local or religious community than other citizens in the state.
3. One has more moral obligation to take care of one's fellow citizens in the state than other people in other states

This Confucian ethics of differentiated and graded moral obligations is not grounded in an assumption that different human individuals possess intrinsic value in different amount or degrees. It rather embodies a relation-relevant morality: for example, I have more moral obligation to take care of my parents than other people's parents simply because they are my parents, not because they possess more intrinsic value than other people's parents. Confucians hold that such differentiated and graded love and obligations reflect "the principle of Heaven and the equity of Earth" (*tianjing diyi*) that needs no justification. They simply manifest the Mandate/Dao of Heaven – it is the command of Heaven that human beings should in such non-egalitarian ways form their relations, exercise their love, and discharge their moral obligations. Nevertheless, Confucian resources still contain some justificatory considerations that we can draw on to defend this non-egalitarian ethics for contemporary discussion.

First, Heaven has invested into the human heart/mind rudimentary moral emotions, especially the feelings of commiseration, shame and dislike, deference and compliance, and right and wrong (Mencius 2A.6). These inherent feelings are already relation-relevant. For example, as Mencius points out, a man naturally loves his brother's newborn baby more than his neighbor's newborn baby (Mencius 3A.5). These innate moral potentials can be cultivated and developed into full-brown moral virtues to treat other people in properly differentiated ways. It is precisely for this purpose of virtue cultivation that the Confucian sages established the Confucian rituals (*li*), a series of familial and social behavior patterns, ceremonies, and conventions, to inform appropriate human relations and guide suitable human interactions. In short, the Confucian life world of performing differentiated and graded moral obligations is in part justifiable through the legitimacy of a Confucian ritual system, because the latter system has a natural and spontaneous root in everyone's inborn moral emotional beginnings invested by Heaven and recognized by the sages to guide and cultivate human virtues in non-egalitarian ways.

Another relevant justificatory reason can be teased out from Confucius' insight into the "rectification of names" (*zhengming*). For Confucius, role names, such as "father," "son," "ruler," and "minister," do not only refer to something in reality; they are operative as well. For example, "father" does not only report mere biological or social facts such as "he is a father"; it is also associated with the obligatory norms of being the father, such as "he ought to be kind to his children." That is, showing special love and taking good care of his young children are the essential obligations of a father that are already implicit in the name of "father." If a father fails to do so, he is not genuinely qualified to be called a father.[7] Similarly, names such as "neighbors," "friends," "fellows," "citizens," "aliens," and "strangers" are all connected with relevant obligations and rituals for human interactions. Although actual moralities operated in various places of the world can be understood as historically, culturally, and conventionally formed particular moral norms which differ from one to another in certain moral content, they commonly manifest such name-relevant and non-egalitarian features. Taken together, they indicate a moral system of differentiated and graded obligations like Confucianism rather than radical egalitarianism. Thus, burden is on radical egalitarians to show why all such non-egalitarian moral systems are fundamentally wrong.

---

[7] For similar justificatory considerations, see Wang 1999 in which he provides a relevant defense of the Confucian duty of filial piety.

Moreover, a Confucian non-egalitarian view of justice or fairness in treating others can be worked out based on the Confucian conception of righteousness (*yi*) to justify unequal love and obligations. For Confucians, righteousness is appropriateness. Is it appropriate that one's moral obligations to other people should be all equal? This requirement would be incongruous with the requirements of basic human relations that Confucianism argues people are naturally falling on: there should be "love between father and son, righteousness between ruler and subject, different functions between husband and wife, precedence of the old over the young, and trustworthiness between friends" (Mencius 3A.4). These different relations embody different moral sentiments and require different rituals for interacting in familial and social contexts, and they necessarily generate unequal moral obligations in treating individuals under different relations. For example, from the Confucian view, given the parent-child relationship held between my parents and myself, it would be squarely unfair for me to exercise no more moral obligation to take care of them than anyone else's parents.

Finally, a justification can be found in the consideration of economic, political, and cultural consequences generated by the egalitarian view of justice and rights. Evidently, any egalitarian universal insurance and similar equal-entitlement-based approaches to the allocation of health-based resources in contemporary social democratic countries would involve both moral and political hazards. Once an equal entitlement is established in society, it will tend to be exploited by many individuals to its utmost. Meanwhile politicians have an incentive to promise, maintain, or enhance welfare benefits, especially health-care benefits, in order to advance their political careers, even if the available resources are insufficient to pay for those benefits. Finally, dramatic demographical changes (including increasing low-birth-rate national states, rocket-rising divorce cases, ever run-up single-parent families, and rapidly aging societies) are not only unbeneficial to the normal development of children but also produce insufficient payers to retirement pensions and ample publically supported egalitarian health-care systems. These factors have combined to raise questions regarding the financial sustainability of the egalitarian view of social justice and rights. Indeed, *national* egalitarian welfare systems have been in crisis in numerous states in today's world, let alone an attempt to building *international* egalitarian welfare systems as some might aspire based on the radical egalitarian interpretation of 14.2. The Confucian moral view of differentiated and graded love and obligation is a useful recipe to prevent the long-term disastrous consequences of radical egalitarianism.

With this Confucian moral system of differentiated and graded moral obligations, we must object to the radical egalitarian interpretation of 14.2. In short, if everyone in the world has a fundamental right to enjoying the equally applied highest possible standard of health all over the world, then firstly (1) we must enforce a one-tier system of health on every country to ensure equality, without allowing any country to provide any better tier of health (than the universally attainable highest standard of health) to its people. This would violate the Confucian conviction that one has more moral obligation to take care of one's fellow citizens in the state than people in other countries. Secondly, following this right, (2) we must require every government to enforce this standard inside the state, without allowing any religious or other communities to offer a better health standard to its members. This would violate the Confucian conviction that one has more moral obligation to take care of those within one's local or religious community than other citizens in the state. Finally, (3) this right requires that everyone equally get access to or enjoy similar health services or goods within any community. This would violate the Confucian conviction that one has more moral obligation to take care of one's family members than others within one's local or religious community.

Confucians would go for the non-egalitarian interpretation of 14.2. The result will inevitably be a multitier system of health both domestically inside a state and internationally among states. Families, communities, and states will each have to be allowed to offer or purchase a better basic tier of health for their respective members.[8] Meanwhile, the support of a multitier system does not preclude, much less deny, a fundamental right that everyone in the contemporary world should have to enjoy the decent minimum standard of health as explained in the non-egalitarian interpretation of 14.2. Although Confucians do not endorse egalitarian love, they do endorse universal love – that is, every human being in the world should be loved and cared in proper ways. This universal love requires the support of establishing a minimum safety net about health for all human beings in the contemporary world, although it rejects the robust idea of equality.

8.5 Conclusion

It is not controversial that progress in science and technology should advance access to quality health care and essential medicines. What is important is to clarify and justify what is a proper right to health in the contemporary world. This essay shows that Confucianism supports that everyone should have a fundamental right to a decent minimum standard of health. This standard should be established internationally, although it is beyond the scope of this essay to define what its specific content should be. However, drawing on the Confucian considerations laid out in the previous sections, I will make two suggestions to conclude this essay. First, this decent minimum standard for global health should not be over ambitious. It may primarily include only public and preventive health measures, including sanitation, vaccination, clean drinking water, basic nutrition and education, and the like. Of course, this standard can gradually be enhanced as science and technology advance. But it should not be set high at this point. Second, major international concerns and efforts should be given to prominent health problems existing in some places of the world, rather than to any egalitarian projects for every place. For example, at the present time, shouldn't the affluent countries in the world be more morally obliged to offer assistance to deal with the largest ever Ebola outbreak that is currently underway in several countries in West Africa?

[8] For a powerful argument for the moral inevitability of a multitier system, see Engelhardt 1996 (esp. pp. 398–404)

References

Beauchamp, T., and J. Childress. 2013. *Principles of biomedical ethics*. New York: Oxford University Press.

Chan, J. 2014. *Confucian perfectionism: A political philosophy for modern times*. Princeton: Princeton University Press.

Engelhardt, H.T., Jr. 1996. *The foundations of bioethics*. 2nd ed. New York: Oxford University Press.

Fan, R. 2010. *Reconstructionist Confucianism: Rethinking morality after the west*. Dordrecht: Springer.

Mencius 2003. *Mencius: A bilingual edition*. Trans. D. C. Lau. Hong Kong: The Chinese University Press.

UNESCO. 2005. *Declaration in bioethics and human rights*. Available at http://portal.unesco.org/en/ev.php-URL_ID=31058&URL_DO=DO_TOPIC&URL_SECTION=201.html. Accessed 20 Dec 2014.

Wang, Q. 1999. The Confucian filial obligation and care for aged parents. In *Confucian bioethics*, ed. Ruiping Fan, 235–256. Dordrecht: Kluwer Academic Publishers.

Chapter 9
Social Responsibility and Health from a Confucian Perspective: A Confucian Response to Ruiping Fan

Roland Chia

Abstract This brief paper is a response to Fan Ruiping's interpretation and evaluation of Article 14 of the UNESCO Universal Declaration on Bioethics and Human Rights from a Confucian perspective. While I agree with the main thrust of Fan's arguments, I focus on a few critical ideas in the paper. I discuss the role of the government and the place of the family and community in Confucian thought. Finally, I examine Fan's understanding of 'minimum standard of health' against Article 14 and concepts of equality and justice.

Keywords Confucian ethics · Social justice · Equality · Healthcare

9.1 Introduction

In this paper Professor Fan attempts to interpret and evaluate Article 14 of the UNESCO Universal Declaration on Bioethics and Human Rights from a Confucian perspective. He argues that the Confucian tradition could fully endorse 14.1, which states that governments as well as society as a whole are responsible for the promotion of health and social flourishing. Fan then proceeds to examine 14.2, pointing out that there are two possible interpretations to the right to health it articulates. Rejecting what he calls the 'radical egalitarian principle', Fan argues in favour of interpreting this clause as making the provision of 'a decent minimum standard of health' obligatory for governments and societies. He maintains that this approach would be consistent with the Confucian tradition, pointing out that its cardinal virtue, *Ren* (humaneness), requires individuals to practise universal but not egalitarian love. In the final section of his paper, Fan briefly describes what 'a decent minimum standard of health' might look like. He suggests that the vision of a decent minimum

R. Chia (✉)
Trinity Theological College, Singapore, Singapore
e-mail: roland@ttc.edu.sg

J. Tham et al. (eds.), *Religious Perspectives on Social Responsibility in Health*, Advancing Global Bioethics 9, https://doi.org/10.1007/978-3-319-71849-1_9

standard should not be internationally ambitious and that prominent health issues, not egalitarian projects, should be accorded priority.

Fan has offered an incisive analysis and interpretation of Article 14 within the political and ethical framework of the Confucian tradition. He has succeeded in showing how the Confucian understanding of society and moral responsibility is not inimical to the moral vision espoused in the UNESCO Declaration, especially with reference to Article 14. I am in total agreement with Fan's reading of this Article in light of Confucian social ethics. I also think that Fan is right to maintain that Confucianism generally does not support an egalitarian approach to health and healthcare and that its approach is more nuanced and perhaps more realistic. In what follows, I would like to take up a few critical ideas in Fan's paper either to expand on them or to throw them into sharper relief.

9.2 The Role of Government

The first important issue that demands closer analysis is the role of government in human flourishing. Article 14 makes it clear that the 'promotion of health and social development' is not only one of the many responsibilities of the government but its 'central purpose'. While this is a responsibility that all sectors of society must share – and we shall look closely at this in the next section – there is a profound sense in which the government must play a significant role. This is because 'health' in this article must be understood in the broadest possible way as wellbeing, making its provision and promotion a complex undertaking. There are two ways in which health in this broad sense could be understood philosophically. The first is the bio-statistical conception, according to which health is understood in terms of a biological state that is typical of a particular species. Thus, according to this approach, 'a healthy organism is an organism where all parts are functioning at their species level'. The second approach is the action-theoretic conception, where health is understood as self-determination, the 'ability to pursue autonomously chosen goals' (UNESCO 2010:20). Although Article 14 appears to favour the second conception of health, both are important in their own distinctive ways. The scope of the task involved in the provision and promotion of health and social development understood in this very broad sense is indicated in the important but by no means comprehensive list provided in14 a-e.

Professor Fan is correct in arguing that 14.1 would pose no real problem for Confucianism. In fact, the Confucian tradition has some very profound stipulations on the role of the king or the government vis-à-vis the people in their charge. Classical Confucian scholars agree that the most basic task of the government is to alleviate suffering and help citizens to pursue material wealth and social wellbeing. In this way, Fan is absolutely right to say that according to the Confucian tradition the government must always exercise benevolent rule over its subjects. This is axiomatically expressed in the proposition 'Heaven did not create the people for the sake of government; Heaven established government for the sake of the people'. In securing material welfare for the people, it is important to note that according to the Confucian tradition, it is not the responsibility of the government to provide for and manage every aspect of their lives. Instead, as Joseph Chan has argued, 'The government's job is rather to provide the conditions in which people can make a living so that they can help their families and relatives if they are in need' (Chan 2003: 237). However, the government's role is undeniable in that its failure to act responsibly could result in hardship. Mencius, for example, believes that poverty is often caused by misrule in the form of heavy taxation, improper distribution of land and indistinct land boundaries (Book I, A:7).

An important virtue that governments must exemplify is *Yi* ('rightness'), which according to Xiao Yang should be understood as justice in Confucian ethics. Now, Confucian scholars including Fan have rightly noted that in the Confucian tradition, there is no direct term for justice. Although in the Confucian literature, there can be found nuanced discourses on social concern and related issues, Fan writes, what is absent is a 'single concept congruent with the Western notion of justice ...' (Ruiping 2003: 143). Scholars, however, maintain that *Yi* comes closest to expressing what may be described as the *sense of justice* in Confucianism. As Alan Fox puts it, '*Yi* as it functions in the Confucian sense is a kind of moral intuition, a sense of right and wrong, which might be described as serving justice, but certainly not constituting it' (Fox 1995: 44). *Yi* can also refer to a fair-mindedness or moral objectivity that refuses to judge others on the basis of prejudices, personal biases and grudges. But, as Xiao Yang has argued, it is not necessary for *Yi* to always mean 'just' for us to come to the conclusion that there is the idea of justice in Confucianism. 'The fact that *sometimes* it behaves like "just"', he writes, 'is enough for proving that there is a concept of justice in ancient China' (Xiao 1977: 531).

Be that as it may, what is interesting is that scholars like Xiao Yang maintain that in Confucianism *Yi* is always exercised in response to concrete situations. In his paper Xiao Yang presents the idea of a 'strict universalist' to bring out the context-dependent way in which the Confucian tradition understands ethics. According to Xiao, a 'strict universalist' would see a course of action as just or unjust in every situation. For him, however, Confucian ethics is profoundly sensitive to what he calls 'the richness and complexity of the particular'. Thus, although Confucianism is not averse to the idea of general rules, it insists that their application depends on the particularities of the situation at hand. Xiao could therefore go so far as to say that 'If someone invariably sticks to a general rule, he does not have *yi*' (Xiao 1977: 540). This has profound implications to the Confucian understanding of distributive justice, as we shall see.

9.3 Family and Community

We turn our attention next to the societal role that is also emphasised in Article 14 of the UNESCO Declaration. The promotion of health and social development is the responsibility not just of governments but also that of 'all sectors of society'. This resonates with the Confucian vision, which, as many scholars have pointed out, is radically social. Using a vivid imagery, the social anthropologist Fei Xiaotong describes Chinese society as 'rings of successive ripples that are propelled outward on the surface when you throw a stone into water. Each individual is the centre of the rings emanating from his social influence. Wherever the ripples reach, affiliations occur' (Fei 1977:211). While Fan is right to focus on the family as the fundamental unit in society, it is the five relationships in the Confucian vision of society that provide the taxonomy of a moral community.[1] The familial context is extremely important, for it is there that self-cultivation begins and develops, and this in turn enables one to act responsibly and compassionately towards others in society.[2] In addition, the Confucian tradition maintains that if everyone plays his or her role properly, there will be peace under Heaven. As Henry Rosemont explains, the five relationships form the essence of human sociality according to the Confucian tradition:

> … there can be no *me* in isolation, to be considered abstractly: I am the totality of roles I live in relation to specific others. I do not *play* or *perform* these roles; I *am* these roles. When they have all been specified I have been defined uniquely, fully and altogether, with no remainder with which to piece together a free, autonomous self (Rosemont 1988: 177).

The Confucian five relationships stress that there can be no 'oxymoronic private citizen', to use Robert Bellah's arresting phrase (Bellah et al. 1985: 271), no split between self and society. The Confucian vision therefore is fundamentally inimical to the Western notion of a contractual society where justice is tied to individual rights (Li 1994:71). Although in Confucianism we have a vision of society that is more communal and relational (as opposed to the contractual), its notion of social responsibility is shaped by the important principle of the 'gradation of love', as Fan has pointed out. This means that although one should be *Ren* to all people, *Ren*'s first and greatest application is still to one's relatives, especially one's parents. 'A person of *Ren* must love first his father and elder brothers', writes Li Chenyang, 'and then, by extension, other people' (Li 1994: 79).

In addition, a Confucian understanding of social responsibility is not based only on the way in which society is imagined and structured. It is also established on the profound virtues whose cultivation it encourages. Confucianism holds that human beings have the basic and natural capacity for sympathy, that is, the ability to identify with the plight of others by putting themselves in their places, so to speak. This is expressed profoundly in the Confucian concept of *shu* ('reciprocity'), which some scholars have described as the principle of reversibility where one imagines

---

[1] Grouped in paradigmatic pairs, the five relationships according to Confucianism cover the whole of society: (1) Ruler and Ministers, (2) Father and Son, (3) Husband and Wife, (4) Elder and Younger, and (5) Friend and Friend.

[2] Thus, according to *The Book of Odes*, 'A young person who is filial and respectful of his elders rarely becomes the kind of person who is inclined to defy his superiors, and there has never been a case of one who is disinclined to defy his superiors stirring up rebellion. The *junzi* applies himself to the roots. "Once the roots are firmly established, the Way will grow". Might we not say that filial piety and respect for elders constitute the root of *Ren*?' (I.2).

oneself in the place of others in order to sympathise with what they are going through. Put differently, *shu* provides the perspectival and phenomenological grounds for appropriate response to the other. Over and above this, there is the virtue of *Ren* – so important in Confucianism – that is variously translated as humaneness, benevolence and even compassion. Mencius made *Ren* the foundation of his ethics when he said that 'No man is devoid of a heart sensitive to the suffering of others ... The heart of compassion is the germ of *Ren*' (2A: 6). *Ren* is the key concept that guides human relations. Thus, Confucius can go so far as to say that '*Ren* is to love others'. *Ren,* therefore, is the perfect virtue, the inner spirit of ritual (*li*). It is therefore pertinent to note that the phrase 'a spirit of brotherhood' was inserted in Article 1 of the UDHR at the request of the Chinese representative C. P. Chang (Chang 2001).

The family-centric character of the Confucian understanding of social responsibility does not imply that one has no moral obligation to meet the needs of those outside of one's social circles. As we have seen, although *Ren* or benevolence is often realised in the context of familial relationships, its application and reach must extend beyond one's immediate social circles. *Ren* also compels us to respond appropriately to strangers in need. Although Confucianism does present a graded view of social responsibility beginning with one's family, the potentially unlimited scope of benevolence suggests that care should also be extended to strangers. In fact 12.5 broadens the idea of the family to include people from the 'Four Seas', that is, all people. Mencius could therefore say: 'A person of *Ren* embraces all in his love' (7A. 46). Therefore it is this whole cluster of virtues, which both the government and society must embody – *yi* ('rightness'), *shu* ("reciprocity'), *Ren* ('humaneness') and *zhou* ('keeping the public interest in mind') – that shapes the Confucian vision of social responsibility. And although it is true to say that in Confucianism the word 'justice' is absent, as Erin Cline has persuasively argued, these ideas 'together reveals a commitment to cultivating a sense of justice in members of society' (Cline 2013: 150).

9.4 Decent Minimum Healthcare

We turn finally, and very briefly, to the discussion on the application of the Confucian vision of society specifically in the provision of healthcare. The assertion in Article 14 that 'the highest attainable standard of health' is a fundamental human right is somewhat problematic and lends itself to different interpretations. Fan is right to reject what he calls a 'radical egalitarianism' that pursues equality for its own sake. Radical or strict egalitarians hold that justice irreducibly requires the quest for a relational ideal that is concerned not just with adequate provision for the worst off because they are doing poorly in absolute terms but rather more fundamentally with how some fare in comparison with others.[3] Critics have mounted a number of

[3] See L. Temkin 1993, 1995; Parfit 1991, 1998.

objections against radical egalitarianism (Lucas 1965, 1977). Once equality is pursued for its own sake, they argue, the wellbeing of the better off must be reduced (levelling down) in order that the wellbeing of the worst off can be improved (levelling up). Apart from the unfeasibility of the proposal, radical or strict egalitarians must also defend their view that equality for its own sake, and not some other moral precept, is indispensable for a theory of justice.

However, equality is an important aspect of justice, and as Amartya Sen has rightly argued (Sen 1992:12–16), every reasonable moral/political theory, including the one that Fan promotes in his paper, is an egalitarian theory. Thus, for Sen, the question is not 'whether equality' but rather 'equality of what?' What is it that should be equally distributed in society for it to be just? And in light of our discussion, what level of healthcare should be made accessible to all, especially to the worst off? In his paper, Fan proposes that everyone has the fundamental right to a 'minimum standard of health'. This approach is to be preferred to the sufficiency criterion proposed by Harry Frankfurt, which is too dependent on personal preferences and desires (Frankfurt 1987). The decent minimum criterion assumes a social consensus on what constitutes a decent level of wellbeing. However, it should be pointed out that with respect to the spirit of Article 14, the decent level of wellbeing must not be given too minimalistic an interpretation. In other words, a decent level of healthcare must not be taken to mean providing just enough to make life marginally tolerable. This in turn suggests that even among countries that offer the 'minimum standard of health', the provision of healthcare for the needy will vary in different contexts, depending on their economic abilities.

References

Bellah, Robert, Richard Madsen, William M. Sullivan, Ann Swidler, and Steven M. Tipton. 1985. *Habits of the heart: Individualism and commitment in American life*. Berkeley: University of California Press.

Chan, Joseph. 2003. Giving priority to the worst off: A Confucian perspective on social welfare. In *Confucianism for the modern world*, ed. Daniel A. Bell and Hahm Chaibonh, 236–253. New York: Cambridge University Press.

Chang, P.C. 2001. Chinese statements during deliberations on the UDHR (1948). In *The Chinese human rights reader: Documents and commentary 1900–2000*, 206–213. Armonk/New York: M.E. Sharpe.

Cline, Erin. 2013. *Confucius, Rawls, and the sense of justice*. New York: Fordham University Press.

Fei, Xiaotong. 1977. Chinese social structure and its values. In *Changing China: Readings in the history of China from the Opium War to the present*, ed. Mason Gentzler, 210–214. New York: Praeger.

Fox, Alan. 1995. The aesthetics of justice. *Legal Studies Forum* 19 (1): 43–56.

Frankfurt, Harry. 1987, October. 'Equality as a moral ideal'. *Ethics* 98: 21–43.

Li Chenyang. 1994. The Confucian concept of Jen and the feminist ethics of care: A comparative study. *Hypatia* 9 (1): 70–89.

Lucas, J.R. 1965. Against equality. *Philosophy* 40: 296–307.

———. 1977. Against equality again. *Philosophy* 52: 255–280.

Parfit, D. 1991. *Equality or priority?* (Lindley Lecture, University of Kansas).

———. 1998. In *Equality and priority' in ideals of equality*, ed. A. Mason, 1–20. London: Wiley-Blackwell.

Rosemont, Henry. 1988. Why take rights seriously? A Confucian critique. In *Human rights and the world's religion*, ed. Leroy S. Rouner, 167–182. Notre Dame: University of Notre Dame.

Ruiping, Fan. 2003. Social justice: Rawlsian or Confucian? In *Comparative approaches to Chinese philosophy*, ed. Bo Mou, 144–168. Hampshire: Ashgate Publishing.

Sen, Amartya. 1992. *Inequality reexamined*. Cambridge, MA: Harvard University Press.

Temkin, L. 1993. *Inequality*. Oxford: Oxford University Press.

———. 1995. Justice and equality: Some questions about scope. *Social Philosophy and Policy* 12: 72–104.

UNESCO. 2010. *On social responsibility and health*. Report of the International Bioethics Committee of UNESCO (IBC), UNESCO.

Yang, Xiao. 1977. Trying to do justice to the concept of justice in Confucian ethics. *Journal of Chinese Philosophy* 24: 521–551.

Chapter 10
Minimun Decency as Ordinary and Proportionate Healthcare Provision: A Christian Response to Ruiping Fan

Peter Au-Yeung

Abstract The endorsement of the first part of Article 14 by Confucianism in that governments should promote health and social developments of its people would find resonance in Catholic social teachings which promotes the exercise of authority as a service. The centrality of the family in society is similarly emphasized by both religions. The right to health may imply a radical egalitarian approach rejected by Confucianism. This is echoed in the Catholic teaching of ordinary versus extraordinary means of therapy. The Catholic vision of equality in the dignity of human persons incorporates a need for service of love (caritas), which expresses itself as love of neighbor. Although Confucianism does not directly address the matter of human solidarity, the codification of obligations between different members of society maintains a structure for maintaining such solidarity. Finally an exploration of ordinary and proportionate means of therapy in the context of current scientific and technological advances may help to set an appropriate minimal decent standard of healthcare.

Keywords Decent minimum healthcare · Human solidarity · Ordinary means · Proportionate means · Radical egalitarianism · Right to health · Service conception of government

Fan presented a critique of parts 1 and 2a of Article 14 of the UNESCO Universal Declaration on Bioethics and Human Rights on the basis of Confucian teachings (especially those expounded in the book of Mencius). His critique contended that Confucianism would endorse the first article that a central purpose of governments would be to promote health and social developments of its people. A large part of the rest of the critique revolved around whether "the highest attainable

P. Au-Yeung (✉)
Diocesan Committee on Bioethics, Catholic Diocese of Hong Kong, Hong Kong, China

J. Tham et al. (eds.), *Religious Perspectives on Social Responsibility in Health*, Advancing Global Bioethics 9, https://doi.org/10.1007/978-3-319-71849-1_10

standard of health" should be interpreted in a radically egalitarian fashion, or taken to mean a highest attainable decent minimum one, and why Confucianism would reject the former and endorse the latter. Having also read Chia's response to Fan, I will also draw upon the observations of the former in assessing how Catholicism would see the response of Confucianism to these elements of Article 14.

10.1 The Promotion of Health and Social Development by Governments

Fan asserted that Confucianism promoted a "service conception" of governments, in which benevolent governments exist for the benefit of people and are justified by their ability to protect and promote the well-being of people, a view echoed and expanded on by Chia in his response. It should also be noted that Fan put the teachings of Mencius in this area in the context of harsh taxation by the warlords of that era, a point also noted by Chia, and this recalls the saying from another Confucian classic that oppressive government is more terrible than tigers (*Book of Rites*, Tang Gong II, 193).[1] The Confucian view of a benevolent government in service to the people would not find any contradiction in Catholic teachings: "Those who exercise authority should do so as a service" (*Catechism of the Catholic Church* (CCC), para 2235).

Mencius taught about the importance for governments to recognize the family for promoting the health and well-being of the people and to protect the integrity, stability, and prosperity of these families. This would certainly find resonance in Catholic teaching: "The importance of the family for the life and well-being of society entails a particular responsibility for society to support and strengthen marriage and the family" (CCC, para 2210). The duties of the political community toward the family, as expounded in paragraph 2211 of the *Catechism of the Catholic Church* (CCC), certainly complement the advice given in the quote of Mencius 1A.7 in Fan's article. The importance of the family in Confucian teachings also relates to the rulers, who are taught first to have "their persons… cultivated, their families… regulated, their states… rightly governed, [before] the whole kingdom was made tranquil and happy" (*Book of Rites*, Da Xue, 2). Not only were the governing elite to pay attention to the importance to the family for the stability of society, they were also enjoined to make sure their own families live in a harmonious and peaceful manner, for he whose family is dysfunctional can never hope to govern the state properly.

[1] The passage recounted Confucius passing by a woman wailing in grief at a tomb. She had lost her husband's father, husband, and son to man-eating tigers but refused to leave the locality because there was no oppressive government there. Hence Confucius taught his followers oppressive government was more terrible than tigers (Legge 1885).

10.2 The Right to Health

Health has been defined in the WHO Constitution as "a state of complete physical, mental and social well-being and not merely the absence of disease or infirmity," and the same document also affirmed that "the enjoyment of the highest attainable standard of health is one of the fundamental rights of every human being." This expansive concept of a right to health can also be found reflected in the various sub-points of Article 14.2, of which we are mainly considering 14.2a. As with Chia, I also note that the *Report of the International Bioethics of UNESCO on Social Responsibility and Health* (IBC report, Paras 22–24) proposed two philosophical concepts of health and discussed their limitations. Leaving aside the nonmedical aspects of such a right, this begs not only the question of attainability, as outlined in paragraph 18 of the same report, but also the question of whether applying an ambitious reading of this right is considered appropriate in the various sociopolitical and cultural contexts around the world. Fan argued that this is not compatible with the Confucian outlook and proceeded to reject the radical egalitarian interpretation of "the highest attainable standard of health" (see next section).

Given the realities of daily life where, even in the most affluent of societies, inequalities of access to top treatments exist, not only by design but also due to serendipity and other factors which cause certain modalities of therapies to be developed in one place rather than another, is it appropriate to consider whether all available technologies (shown to be of benefit to patients) should be universally and equally distributed or that everybody in the world be given equal access, no matter how?

The Catholic approach would not see this as appropriate. The CCC teaches that "Life and physical health are precious gifts entrusted to us by God. We must take reasonable care of them" (CCC, para 2288), but "If morality requires respect for the life of the body, it does not make it an absolute value" (CCC, para 2289). Traditional Catholic medical ethics speak of ordinary and extraordinary means of therapy (which can cause confusion due to conflation between the philosophical usage with the everyday usage of the words ordinary and extraordinary); a clearer exposition of the concept can be found in the 1980 Declaration on Euthanasia which discussed it in terms of proportionate and disproportionate means (Declaration on Euthanasia, Congregation for the Doctrine of Faith 1980, Section IV, Due Proportion in the Use of Remedies). In this way, Catholic medical ethics would not support this type of right to health.

10.3 Rejection of the Radical Egalitarian Approach

It may be argued that the formal rejection of the radical egalitarian interpretation of Art.14.2 by Fan is unnecessary; paragraph 39 of the IBC report seemed to have rejected this interpretation when it noted two different levels of governmental responsibility: "Governments have a first and primary obligation to the people for whom they are directly responsible. Beyond this duty, there is an obligation to help other countries, which is increased in so far as more resources are available and the governments abroad lack the means to protect the health of their people."

This would be very similar to the "Confucian moral system of differentiated and graded moral obligations" Fan defended in section IV of his paper and with which Chia agreed in the final "Decent Minimum Healthcare" section in his response to Fan.

Radical egalitarianism also does not form part of the Christian vision of mankind; another kind of equality is envisaged. "The equality of men concerns their dignity as persons and the rights that flow from it" (CCC, para 1945; also see para 1935). But there also exist "differences among persons belong to God's plan, who wills that we should need one another. These differences should encourage charity" (CCC, para 1946; also see paras 1936–7). The radical egalitarian reading of "the enjoyment of the highest attainable standard of health [being] one of the fundamental rights of every human being" would also require a Marxist-type pledge "to build a just social order in which all receive their share of the world's goods and no longer have to depend on charity"(Deus Caritas Est, Benedict XVI 2005, para 26). But whether that is achievable or not, there can be "no ordering of the State so just that it can eliminate the need for a service of love. … The State which would provide everything, absorbing everything into itself, would ultimately become a mere bureaucracy incapable of guaranteeing the very thing which the suffering person—every person—needs: namely, loving personal concern" (Deus Caritas Est, Benedict XVI 2005, para 28(b)). In any case, "Love - *caritas*- will always prove necessary, even in the most just society. …There will always be suffering which cries out for consolation and help. There will always be loneliness. There will always be situations of material need where help in the form of concrete love of neighbour is indispensable." (IBC Report, Paras 22–24)

10.3.1 Charity and Human Solidarity

Love of neighbor has a special place in the Christian vision – in answering a question on the laws and commandments, Jesus taught that second to loving God, "You shall love your neighbour as yourself" (Mt 22:39). In a separate account (Lk 10:26–8), Jesus taught the same to a lawyer and further explained who a neighbor really is, by telling the parable of the Good Samaritan. This latter is further explored by Durante in the Christianity session and also by Heyd in response, so this will not be further dealt with here. A further indication of the importance God places on loving the needy comes in the passage on the judgment of the nations (Mt 25:31–46), which can be summarized as "Truly, I say to you, as you did it to one of the least of these my brethren, you did it to me" (Mt 25:40).

Fan did not directly address the issue of human solidarity but hints at it in Section IV, when he discussed the rectification of names and its importance in ordering obligations in different societal relationships in Confucianism. Chia expanded on this in the Family and Community section with a discussion of the Confucian Five Relationships. What these already illustrate is that Confucianism regulates human relationships and thus codifies the proper level of obligations between different members of the society and hence in a way gives structure to human solidarity in ancient Chinese society. In one sense, this is different to the Christian view, because the Good Samaritan is really an alien, and outsider, but then "[Christians] reside in their own nations, but as resident aliens. They participate in all things as citizens and endure all things as foreigners...." (Quoted in CCC 2240) The outsider, the Good Samaritan, does what is over and above what is required of him, and the concept of supererogation will be further elaborated in Heyd's paper.

10.3.2 Decent Minimum Healthcare

Having rejected a radical egalitarian interpretation of the right to health, Fan argued that there should be a decent minimum standard of health globally, which need not be internationally ambitious but which should be enhanced as science and technology advanced. Chia agreed but cautioned that this should not be set so low as to make life marginally tolerable right from the very beginning, and this latter warning against marginally tolerable health recalled Parfit's Repugnant Conclusion, in his critique of utilitarianism. The Catholic view of ordinary or proportionate and hence morally obligatory interventions may offer one way to ensure that a decent minimum level be set above the marginally tolerable level (cf. CCC 2278–9). This is not unlike the critical level principles which some use to counter the Repugnant Conclusion (Arrhenius et al. 2014). Yet like Chia's last observation, this can take into account the context of the patients in terms of the different socioeconomic they face in different settings around the world, as well as urge the more affluent countries to do more to raise the level of such a decent minimum standard in those societies that are less well-off.

References

Arrhenius, Gustaf, Jesper Ryberg, and Torbjörn Tännsjö. 2014. "*The repugnant conclusion,*" *The Stanford encyclopedia of philosophy* (Spring 2014 Edition), ed. Edward N. Zalta. http://plato.stanford.edu/archives/spr2014/entries/repugnant-conclusion/.

Catechism of the Catholic Church, Libreria Editrice Vaticana, Citta del Vaticano. 1993.

Declaration on Euthanasia, Congregation for the Doctrine of Faith. 1980.

Deus Caritas Est, Benedict XVI. 2005.

The Book of Rites (English version: "Sacred books of the East, volume 28, part 4: The Li Ki," James Legge, 1885).

UNESCO. 2010. *The Report of the International Bioethics of UNESCO on Social Responsibility and Health.*

Part IV
Hinduism

Chapter 11
Hinduism and Social Responsibility

Vasantha Muthuswamy

Abstract Hinduism, known initially as *Sanatana Dharma* (eternal or timeless religion), is more of a "way of life" than an ideology; it is a lived religion, not simply a cognitive one. The Hindu way of life is a science of salvation to liberate the human beings from the cycle of births and rebirths so that the individual soul reunites with the universal soul. Vedanta provides the highest spiritual knowledge and the code of conduct and spiritual values. Vedantic philosophy developed these values to anticipate the socio-economic problems fully. It prescribes the *Dharma*, the righteous duty of each, and *Karma*, the line of selfless action to be undertaken without expecting any return towards individuals, family, society, nation and the Universe. The concepts of the four goals and five debts everyone is born with, as per Hindu belief, emphasise our interpersonal duties and social responsibilities in this life. They provide each person with a moral compass to guide one's life by cultivating family and social values to fulfil the social responsibilities towards the weak and the needy. Thus, Hindu ethics leads eventually to self-realisation or liberation from the cycle of birth and death, *moksha*, through service to humanity. Hindu philosophers encourage the performance of duties not with a sense of compulsion but through love which springs from the perception of God in all living beings. We examine the concept of UNESCO Declaration as applicable to human rights, fundamental freedom, and human dignity in this paper through the Hindu philosophical approach from time immemorial. They are the individual and societal duties and responsibilities that form the current guiding principles of Constitution of India for protection of the fundamental rights of its citizens. We also elaborate on how the Hindu tenets of *Dharma* and *Karma* execute their social responsibility in religious institutions and modern corporate houses in promoting Health for All.

Keywords Hinduism · Dharma · Karma · Moksha · Corporate social responsibility

V. Muthuswamy (✉)
Forum for Ethics Review Committees in India, Mumbai, India

J. Tham et al. (eds.), *Religious Perspectives on Social Responsibility in Health*, Advancing Global Bioethics 9, https://doi.org/10.1007/978-3-319-71849-1_11

Hinduism, originally known as *Sanathana Dharma* (eternal or timeless religion) is more of a "way of life" than an ideology; it is a lived religion not simply a cognitive one. It is very difficult to separate living and religion in Hinduism. The word "Hinduism" itself is not found in any of the ancient Indian scriptures. It was the term used by the Persian invaders to designate the people living east of river *Sindhu* and the Sanskrit word pronounced as Hindu by the Persians, and hence, the religion of the people living in this area came to be known as Hinduism. Being a Hindu does not mean having a particular belief system. Basically, the whole culture is oriented towards realising one's full potential. There is no particular God or ideology that one can call as the Hindu way of life. You can be a Hindu irrespective of whether you worship a man-god or a woman-god and whether you worship a cow or a monkey or a tree. People worship whatever aspect of life they relate to most, and that is perfectly fine. One can still be a Hindu without worshipping anything. Hindu was never an "ism", and the attempt to organise it as a religion is still not successful because the Hindu way of life which is referred to as *Sanatana Dharma* or universal law is all inclusive in nature and does not exclude anything. The Hindu way of life is not an organised belief system but a science of salvation to liberate the human beings from the cycle of births and rebirths so that the individual soul reunites with the universal soul.

The source of Hinduism is *Vedanta*. Vedanta is one of the world's most ancient spiritual philosophies and one of its broadest, based on the Vedas, the sacred scriptures of India. It is the philosophical foundation of Hinduism. "Vedanta" is a combination of two words: "Veda" which means "knowledge" and "anta" which means "the end of" or "the goal of". "Knowledge" here means the knowledge of God as well as the knowledge of our own divine nature. Vedanta for the Hindus is considered to be the highest spiritual knowledge containing self-knowledge and the knowledge about the Absolute Truth. All forms of Vedanta are drawn primarily from Vedic literature comprising of four Vedas (*Rig*, *Yajur*, *Sama and Atharva)*, the two epics – the *Mahabharatha* by *Vyasa* and the *Ramayana* by *Valmiki – the puranas*, *samritis* and the commentaries of great teachers through centuries (Duties and Responsibilities, the Hindu Social Framework. http://artstudio.co.za/Duties_and_Responsibilities_The_Hindu_Social_Framework.html.). As a way of life, it accepts and assimilates varieties of faiths eloquently expressed by the Vedic seers in the phrase, "Truth is one; sages call it by various names". This is the foundational philosophy of Vedanta and is the greatest pronouncement in the *Rigveda* (Swami Ranganathananda 1971).

Of all the scriptures, the *Bhagavad Gita* (a part of Mahabharatha) is the most beloved for all Hindus like the Qur'an for Muslims and the Gospels for the Christians. It is supposed to have been narrated by Lord Krishna during the Kurukshetra war in 3102 BC, which corresponds to 1700 years before Moses, 2500 years before Buddha, 3000 years before Jesus and 3800 years before Prophet Mohammad (Hindupaedia. The Hindu Encyclopaedia. http://hindupedia.com/en/Main_Page. Accessed 5 June 2014.). The code of conduct and the spiritual values that were developed by Hindu seers on the basis of Vedantic philosophy seem to have fully anticipated the socio-economic problems that civilisation has had subsequently to face. The profit-motive and the civic right of private competition were definitely discarded in what was laid down as the Vedantin's way of life. This, as clearly set out in the Bhagavad Gita, is that men must fulfil social duty and work according to capacity and not for profit. We are now told by social and economic reformers that the State should see to it that men and women work without aiming at personal gain and with an eye only to the welfare of the community. And this is just what the Bhagavad Gita laid down. The way of life taught in this living spring of Hindu ethics is based expressly on the equal dignity and sacredness of every form of labour that falls to one's lot. All work, it reiterates with solemn emphasis, should be done honestly and disinterestedly for *lokasangraha* – welfare of the community – and not for the satisfaction of personal desires. Indeed, the Gita lays down in a unique manner the whole socialist doctrine by characterising work as a religious offering in the truest sense. The performance of one's allotted task is specifically described in the Gita as an authorised and accepted form of worship:

The Bhagavad Gita starts with the word *Dharma* which is an important concept in the Vedanta philosophy in India and means one's *righteous duty*. The Sanskrit word for *Dharma* has no direct translation into English. It can be thought of as the word that means righteousness in thought, word and action. Another interpretation of the word would be "collection of natural and universal laws that uphold, sustain or uplift". There are many interpretations to this word in different texts:

> It is very difficult to define Dharma. It has been explained to be that which helps the upliftment of living beings meaning that which ensures the welfare of living beings. – *Mahabharatha* (Bhishma) (Mahabharatha, Shanti Parva (109-9-11))
>
> Dharma sustains society, it maintains social order, it ensures well being and [the] progress of humanity. – *Karna parva* (Lord Krishna) (Mahabharatha, Karna Parva (69.58))
>
> Dharma is that which is indicated by the Vedas as conducive to highest good. – *Purva Mimamsa* (Rishi Jamini) (Purva Mimamsa, Rishi Jamini (1-2))
>
> Dharma embraces every type of righteous conduct, covering every aspect of life essential for the sustenance and welfare of the individual and the society. – *Parashara Smriti* (Madhavacharya) (Parashara Smriti, Madhvacharya)

The earliest import of the word *Dharma* arises from the ancient Vedic idea of *Ritam* or Cosmic order, which are physical laws that govern the motion of stars, suns, planets, satellites, asteroids and other physical bodies. In addition to this physical Universe, the word is then applied to Human life as stated by Lord Krishna that *Dharma* sustains society, maintains social order and ensures well-being and progress of humanity Pattanaik, Devdutt (2010). *Dharma* as a social order has given an elevated place in the consciousness enshrined through numerous stories and *puranas*, epics and tales in order to make it accessible to all ranging from uneducated to the rulers, from the sages to merchants, irrespective of their station and role in society. This ensures that *Dharma*, the law of right living and conduct, is enshrined and passed down from generation to generation. Ramayana and Mahabharata are known as *Ithihasa* and are distinct from *Vedas* and *Puranas*. *Ithihasa* is the struggle of man to uphold dharma in his pursuit of perfection and divinity. The *Vedas*, by contrast, list the principles governing life in an abstract manner, while the *Puranas* embody these principles in various divine beings and tell stories how God cyclically creates and destroys the world (Arvind Sharma 2002).

Dharma is the first of the four goals of human life defined by Hinduism which are called the *purusharthas* (*purusha* means a human being or God and *artha* means an objective goal). These four goals, namely, *Dharma*, *Artha*, *Kama and Moksha*, create a social framework within which one can live peacefully and gradually evolve spiritually. The root of the word *Dharma* comes from the word *dhri* which means to uphold or maintain. *Dharma* forms the foundation for the other three goals – *Artha* (*success*, material wealth or economic activities), *Kama* (desire or pleasurable pursuits) and *Moksha* (spiritual perfection leading to liberation). *Upanishads* (800–300 BC) are a collective of texts that contain the central philosophical thoughts of Hinduism. There are more than 200 of these, out of which a little over dozen are the main ones. The major focus of *Upanishadic* thinkers was the idea that the individual soul, called *atman* was a separated part of *Brahman*, the Universal soul. The separated *atman* after achieving *moksha* or liberation from life and rebirths will be united again with *Brahman*. The famous phrase of Chandoga Upanishad "tat twam asi" meaning "you (the individual soul) are that (the Universal soul)" is the principle teaching of Hinduism (Wangu, Madhu Bazaz 2009). In order to understand how Hindus conceptualise their struggle to achieve the ultimate goal of *moksha*, one must first understand the two principal elements of *samsara* and *karma*.

The word Hindus use for the process of going from one rebirth to another is "samsara". Samsara means "to cycle or go around". Samsara represents the cycle of life, death and rebirth in which a person carries his or her own karma. Each life cycle presents an opportunity for balance.

An individual may experience effects from past lives, although the circumstances may be totally different. Hindus believe that a person's worldly status depends upon actions in a past life. However, it may be many lifetimes within the wheel of life before *moksha* is achieved. The basic idea behind *karma*, which is an important aspect of the cycle of samsara, is that good actions produce good results, and bad actions produce bad ones. While on the wheel of samsara, each soul is either exhausting past *karma* or accumulating fresh *karma*. Negative *karma* accumulated from previous lifetimes may show up in this one as hardships or misfortunes that are otherwise unexplainable. Similarly, positive *karma* may produce fortuitous circumstances or meetings with teachers that advance the soul in its spiritual practice. Liberation occurs when the soul exhausts its outstanding *karma* and becomes detached enough to avoid creating more. Thus, Hinduism gives opportunity to individuals to perfect themselves through love and action at different rebirths in different forms to achieve the final goal of reunion with the Universal soul and liberation from the cycle of birth and rebirths.

Hinduism describes *dharma* as the natural universal laws whose observance enables humans to be contented and happy and to save himself from degradation and suffering. *Dharma* is the moral law combined with spiritual discipline that guides one's life. Hindus consider *dharma* the very foundation of life. It means "that which holds" the people of this world and the whole creation. *Dharma* is the "law of being" without which things cannot exist.

Hindu philosophers emphasised personal ethics over social ethics. Their argument was that since society consisted of individuals, if individuals were virtuous, social welfare would follow as a matter of course. Moreover, the general moral tone was very high in the ancient Hindu society, where everybody was expected to do his appropriate duties, which included, among other things, rendering help to one's less fortunate fellow beings. In Hinduism, dharmic activities correspond to individual, social, political and professional actions, all of which are based on four virtues – truth, *ahimsa* (non-violence), morality and duty. The purpose of *purushartas* is to ensure that people would not neglect their obligatory duties by becoming obsessed with particular desires that may lead to moral and social degeneration of values. To further ensure a stable social framework, Hinduism introduces the concept of debts or indebtedness. According to *Taitriya Samhita* (6.3.10.5) and *Shatpath Brahman* (1.7.2.1–6), a child is born with five debts to repay in his/her lifetime:

(a) *Deva rin* – indebtedness to Gods/Deities for having access to God's creations
(b) *Rishi rin* – indebtedness to ancient sages and intellectuals/teachers for the knowledge acquired
(c) *Pitri rin* – indebtedness to ancestors for being responsible for one's birth
(d) *Nri rin or Manushya rin* – indebtedness to humanity which can be repaid by treating them with respect and by helping them
(e) *Bhuta rin* – indebtedness to plants, animals and nature or environment which can be repaid by protecting them

One can see that the concept of "service" has its roots in the ancient Hindu tradition of the fivefold debt system. The concepts of the four goals and the five debts generate an awareness of one's interpersonal duties and social responsibilities in this lifetime and provide each person with a moral compass with which to guide one's life by fostering the cultivation of family and social values.

Dharma includes the golden rule "Do unto others as others would do unto you" and therefore the set of all possible values related to specific way of life, value systems and attitudes ensuring ethical behaviour in the society. Thus, *Dharma* is expressed as the duties and responsibilities of an individual or community that ensures the harmony and balance in society as a whole, in terms of its interrelations and its dependencies (Duties and Responsibilities, the Hindu Social Framework. http://art-studio.co.za/Duties_and_Responsibilities_The_Hindu_Social_Framework.html). So *Dharma* upholds, sustains and uplifts all the various constituents of the Universe, who/which are woven together in a common interdependent existence, for example, parents vis-a-vis children, state vis-a-vis citizens, society vis-a-vis environment, etc.

11.1 *Dharma* as a Social Responsibility and Service to Community

Religion and philosophy play a significant role in the development of human values and behaviour which in turn also influences business and corporate management. Research also suggests that the encouragement of religious principles in business can lead to benefits in the area of creativity, honesty and trust, personal fulfilment and social commitment which will ultimately lead to increased corporate and business performance. *Sarva loka hitam* in Vedanta refers to "well being of all stakeholders in society". The *Bhagavad Gita* also stresses that duty needs to be done without any attachment and only those who do their duty without attachment will attain the supreme goal of liberation (BG 3.19). The Law of *karma* (cause and effect) is expected to motivate the business people to carry out their duty towards humanity. The Vedanta also promotes the concept of *Nishkama karma* – a perspective on action and decision-making that emphasises performing one's duties without attachment to its fruits thereof – as both the action and its fruits are offered to the divine. Hence, corporate social responsibility (CSR) is a contribution to the welfare of the society to be carried out as a duty and without expectation of any returns from the society. Kautilya sastra (fourth century BC) states "in the happiness of his subjects lies the happiness of the king; and in their welfare his welfare; the king shall not consider as good that which pleases him but treat as beneficial to him whatever pleases his subjects – welfare of the many and the happiness of the many". This is also in line with the popular vedic prayer, *loka samastha sukino bhavantha*, meaning "all may be happy and healthy" (Muniapan and Satpathy 2013). Mahatma Gandhi explains with the authority of his personal experience: "I mean that you must not worry whether the desired result follows your action or not, so long as the motive is pure, your means are correct". By renunciation of the fruit, Gandhi warns "in no way means indifference to the result. In regard to every action, one should know the result that is expected to follow, the means thereto, and the capacity for it. He, who, being thus equipped, is without desire for the result and is yet wholly engrossed in the due fulfillment of the task before him, is said to have renounced the fruits of his action" (Easwaran 2010). Thus Hinduism greatly supports corporate social responsibility as it encourages people to be responsible and take on their rightful duties, which would be the first step to being socially responsible. The next step would be to perform duties that exceed compliance by extending love, fairness and good work ethic. One can then assume that corporate social responsibility leads to moksha, or conversely, the desire to attain moksha encourages corporate social responsibility.

Generally human endeavours involve the pursuit of other goals in life, namely, *Artha* and *Kama*, respectively, meaning material wealth (*artha*) and pleasure in personal and professional life (*kama*). However, when this is extended to the wider possibility of subordinating personal interests in the interest of ensuring the well-being of others – family, neighbourhood, village, city or state or a specific community or even country or the world – their lives take on a wider dimension. A commitment to pursue the greater good clearly sustains, maintains, upholds and uplifts the prevailing social order, especially the community that is the beneficiary. We have seen that the concept of "service" has its roots in the ancient Hindu tradition of the fivefold debt system. The concept of the four ends and five debts generates awareness of one's duties and responsibilities, provides moral and ethical direction to life, encourages family values and helps one to organise life for individual accomplishments.

Dharma is the power that maintains society; it makes the grass grow and the sun shine and makes us moral people or rather gives humans the opportunity to act virtuously. But acting virtuously does not mean precisely the same for everyone;

different people have different obligations and duties according to their age, gender and social position. *Dharma* is universal, but it is also particular and operates within concrete circumstances. Each person therefore has their own *dharma* known as *sva-dharma*. What is correct for a woman might not be for a man or what is correct for an adult might not be for a child. The human life according to Hinduism goes through four stages of almost 25 years each. The role and responsibilities/duties at each one of this stage are well prescribed. These are known as the *Ashramas*, viz., *Brahmachari* (student), *Grihasta* (householder), *Vanaprastha* (forest dweller) and *Sannyasa* (renounced or retired), and the *dharma* of each stage is different. These are said to be representing preparedness, production, service and renouncement (Bose 1998).

Dharma is therefore inherent to any community service and is a life of selfless service, of sacrifice and contribution. Making a difference in society at large honours and fulfils the human spirit and brings to life the true possibility and potential of human existence. Every epic and *Purana* thus extol *Dharma*. In Ramayana, this idea is personified as Lord Rama – the embodiment of *Dharma*.

The Hindu philosophers, without encouraging the illusion that a perfect society could be created, always exhorted people to promote social welfare as a part of spiritual discipline. We must do good to others, because by means of selfless action, we can purify our hearts and transcend the relative world of good and evil. Social service has only an instrumental, not an ultimate, value.

Dharma is classified into many subsets depending on application area and to whom or what is being discussed. For every human being on earth, the following *Dharmas* are applicable:

- *Dharma* of an individual – *Vyakti Dharma*
- Family *Dharma* – *Kutumbha Dharma*
- Societal *Dharma* – *Samaj Dharma*
- National *Dharma* – *Rashtriya Dharma*
- *Dharma* of Mankind – *Manav Dharma/Manushya Dharma*
- Professional *Dharma* – *Varna Dharma*
- Exceptional situation *Dharma* – *Apad Dharma*
- *Dharma* at different ages – *Yuga Dharma*
- *Dharma* at different stages of life – *Ashrama Dharma*

In addition to the above, Mahatma Gandhi propagated *Ahimsa Paroma Dharma* that non-violence is the topmost *dharma* for all, and this is appreciated all over the world (Gandhi 1958). *Thus one can realise that there are ten important duties enjoined upon Hindus by Vedas and Dharmasutras (book of duties).* These are:

- Duties towards self
- Duties towards Gods
- Duties towards ancestors
- Duties towards progeny
- Duties towards fellow human beings
- Duty towards other living beings

- Duty towards society
- Moral duties to practice virtue and to uphold divine laws
- Professional duties and
- Duty towards other faiths

Samanya dharma entails the generic duties to be practised by all. Goodness and service to others is not the property of any one class, creed, sect or community. Hence, every man should possess this virtue. *Samanya dharma* enjoins individuals to act in ways commensurate with non-violence (*ahimsa*), truthfulness (*satya*), non-stealing (*asteya*), purity (*sauca*) and sensorial restraint and considered universally valid. However, there are special duties to be followed or practised by particular caste groups or men in particular stages of life.

The Vishesha *dharma* addresses specific groups that are particularly vulnerable such as women, children, elderly, physically challenged, mentally disabled, etc., the duties towards which are considered to be context specific. As the Manu Smriti reminds us (185.1.110), these are further classified into *Varnashrama dharma* (caste specific), *Stree dharma* (women specific), *Raja dharma* (duties of rulers) and *Yati dharma* (duty of ascetics) to name a few. Feeding the poor and hungry, clothing the naked, giving shelter to homeless and comforting the distressed etc. are all forms of *Manav* or *Manushya dharma* (duty to humanity).

By performing activities of kindness and sympathy, man cultivates cosmic love. Hindu philosophers encourage the performance of duties, but they exhort men to perform them not from a sense of compulsion but through love. Unless a man is inspired by love, he cannot cheerfully perform his duty at home, in the office, in the factory or on the battlefield. This love is not, however, sentimentality but springs from the perception of God in all living beings. Work done under the impulsion of duty deepens a man's attachment to the world, but when performed through love, it brings him nearer to freedom. He learns that he can be happy only by making others happy, by serving others, by removing the suffering of others and by sharing what he has with others. These daily sacrifices establish his relationship with others in the society – his superiors, his equals and his subordinates. His whole life must be one of sacrifices and duties. Only then will he realise the supreme bliss of the eternal to free himself from the cycle of births and deaths to attain liberation and immortality.

While rights and responsibilities are two sides of the same coin, *Hindu Dharma* emphasises responsibilities and duties over rights, while other civilisations emphasise rights. It was clearly made to realise that when responsibilities and duties are fulfilled by one, the others receive their rights. When people live in conformity with *Dharma*, the individual rights of all others are naturally granted. A culture that insists on rights over duties promotes competition and a constant struggle for attaining rights. At first glance, the emphasis in the *Hindu* and other *dharmic* traditions may appear to be primarily concerned with inner self-realisation. However, all are encouraged to engage in community service as a transformative practice, *sadhana*, without personal recognition or publicity. The underlying Vedic philosophy, often quoted by Mahatma Gandhi, maintains that "Service to man is service to God" (*Nar seva, Narayana seva*) (Viswanathan 2008) and is enshrined in *Bhagvad Gita* – 17–20:

> Service which is given without consideration of anything in return, at the right place and right time to one that is qualified, with the feeling that it is one's duty, is regarded as the nature of goodness.

Growing up as a Hindu, religion is always part of one's life in the sense that there are always rituals to learn about, ceremonies to perform, classes to attend, epics to read and recite, temples to visit and stories and plays to watch and listen to. The main lesson the religion offers to learn through all these is one of dedication and detachment – intertwining the concepts of *Dharma and Karma* to work hard, to do your best at everything you do, but to remain unattached to the results of those actions, because these are beyond one's control. The common interpretations vacillate between fatal acceptance of one's fate to live the life they lead (karma theory)and the responsibility of others to perform their duties to alleviate the suffering of others (dharma theory). This brings out clearly that service to humanity is an important responsibility of all human beings as per hindu philosophy.

In the teachings of Mahatma Gandhi and later interpreted by Martin Luther King Jr., Gandhi said that "the whole gamut of man's activities – constitutes an indivisible whole. You cannot divide life, social, political, economic, and purely religious into water tight compartments" "Although not a Hindu by Faith, Dr. Rev. Martin Luther King Jr. studied Gandhian philosophy and spoke about the "inter-related structure of reality" where "whatever affects one directly, affects all indirectly". His words further delineate a Hindu theology of providing a powerful source for united front in the struggle for social justice whereby *dharma* states that it is the duty of all to do what is right and good for all. So here is the quintessential Hindu way of understanding one's own *dharmic* path through self enquiry.

Service or *Seva* is the cornerstone of the Vedantic Hindu philosophy of Yoga as described in Bhagvad Gita and other sacred literature (Bhargava. The importance of seva and social justice for inner transformation," in https://www.huffingtonpost.com/anju-bhargava/hindu-aevaand-social-jus_b_824360.html). Ethical actions calculated to promote social welfare is enjoined upon all who are identified with the world and conscious of their social responsibilities. The observance of social ethics, in a large measure, preserved Hindu society when various outside forces threatened to destroy it. The neglect of social ethics, on the other hand, has undermined its vitality. However, Hindu philosophers emphasised personal responsibilities over social responsibilities. Second, the general moral tone was so high in the ancient Hindu society to the extent that everyone was expected to carry out his appropriate duties, which included among other things, rendering help to one's less fortunate fellow beings. The Hindu philosophers, without encouraging the illusion that a perfect society could be created, always exhorted people to promote social welfare as a part of spiritual discipline. One must do good to others, because through selfless action one can purify one's heart and transcend the relative world of good and evil. The performance of duties was encouraged not from a sense of compulsion but through love. This love springs from the perception of God in all living beings. Work done under the compulsion of duty deepens the attachment to the world, but when performed through love, it brings him nearer to liberation.

The healthy social environment created by objective ethics provides men with an opportunity to cultivate the more important subjective ethics. The *Bhagavad Gita* says, "Let a man lift himself from his own self, for he himself is his friend and he himself is his enemy" (Swami Nikhilananda 2014). Human society is like a huge machine. The individuals and communities are like its parts. If the parts are weak or broken, the machine will not work. Every individual should perform his duties efficiently for the welfare of the society. Peace and order will prevail in society only when all its members do their duties efficiently. A Hindu believes that in loving others, he loves himself, and in injuring others, he injures himself. Through the practice of cosmic love, he feels that all bodies are his, all hands are his and all feet are his and that the whole world is his home (*Vasudeva Kutumbakam*). Thus, Hindu ethics leads eventually to *self-realisation through service to humanity*.

There are several Hindu religious groups which have in their mandate service to suffering humanity for betterment of health status as a duty prescribed by the religion. There are hundreds of such organisations all over the country catering to different population groups. Vishwa Hindu parishad (VHP) is a Hindu right wing organisation based on the ideology of *Hindutva*. Its main objective is to organise, consolidate and serve the Hindu society and to protect the *Hindu Dharma*. The members of VHP train people in villages to provide primary health-care and referral services. They also conduct regular medical check-up camps in different parts of the country (Whaling 2013).

Another well-recognised religious movement linked to Hindu tradition is the *Brahma Kumaris* (BK) World Spiritual University founded in the 1930s which is noted for its female leadership. From the beginning, the organisation is focussing its activities in the area of education and free health-care services to one of the most impoverished areas in western India. In more recent years, the BK's expansion in service has led to a greater participation in more main stream community health services in the form of Charitable village outreach programmes and establishing Global hospital and Research centre in the State of Rajasthan. A rural health care and eye hospital is set up at the base of Mount Abu in 2004. The BKs won the UNESCO special award for promotion of the "Culture of Peace" project in 2000 (VHP's Social service activities 2011). Ruth Bryod Sharone, the co-chair of the Interfaith Ambassadors for the Parliament of the World Religions (IAPWR) and journalist, wrote: "BKs are well known for their charitable acts, especially in Mount Abu where they have established themselves as teachers and healers. A modern hospital provides low cost health care not only for the BK community but the entire population of Rajasthan. I visited one of the villages and met the 'miracle children' who would have died or have been severely crippled from malnutrition and sickness were it not for the BKs education" ("A Timeless Woman with a Timely Message." The Interfaith observer. http://www.theinterfaithobserver.org/journal-articles/2011/12/5/a-timelesswoman-with-a-timely-message.html. Accessed 24 July 2014).

The Sankara Mutt, a leading Brahmin organisation in South India, has been doing yeomen services combining religion with social welfare. Their endeavour is upliftment of many hospitals and health-care organisations which cross subsidise each other. The most famous of all these is the Sankara Eye Centre (*Sankar Nethralaya*) which gives eye care to all strata of society. Half of its surgeries are done free of cost. This public charity network of eye hospitals run by Sri Kanchi Kamakoti Medical trust under the Mutt has reached out to 20 million rural poor and performed over 200,000 free eye surgeries.

A historic Hindu spiritual and service fair was organised in the southern city of Chennai in 2004 where more than 40 Hindu religious and spiritual organisations which have been silently working in vulnerable populations for many decades avoiding publicity and media glare showcased their activities for the first time. Some of these well-known organisations are Arya Samaj, AIM for Seva, Brahmakumaris, Art of Living, Chinmaya Mission, Divine Life Society, Vidya Bharati, Seva Bharathi, Sri Kanchi Kamakoti Peetam, Patanjali Yoga Peeth, Ramakrishna Mutt, Ekal Vidyalaya, Swami Vivekananda Kendra, Mata Amruthanandamayi Mutt, Oneness University, Hindu Dharma Acharya Sabha, Sathya Sai Foundation, Ramakrishna Mission, etc. to name a few.

In addition to the above, there are many other organisations with branches abroad, which provide humanitarian activities for providing education, training and health care to those in need. These activities are extension of the idea of "Serve people to serve God". The activities reflect a shift in the approach from *Sadhana* (Penance) to "Service" putting human values before religious concepts. Hinduism has been a nondoctrinaire, non-formalistic umbrella religion which has absorbed ideas from various local Indian cults over a period of time. The Hindu religious organisations have made significant contributions to society by paying more attention to religious values than religious rituals, by re-explaining religious concepts in a way that suits the new age and the newly emerged social conditions and by demonstrating universal values through social programmes and social services (Sivananda 1999).

Sri Sathya Sai Medical Trust at Bengaluru, India, with branches in and outside India sums up the guiding principles for health care that are applicable for all such institutions:

- Medical care is the right of every individual regardless of caste, creed, colour or nationality.
- The act of healing should be restored to its pristine sanctity, which can be done only when it ceases to be a commodity bought and sold in the market.
- The human agents of the delivery system of health care, be the health administrators, doctors, nurses or technicians, should be perfectly "human" in thought, word and deed. They should demonstrate the five human values – truth, righteousness, peace, love and non-violence – in every act of theirs.
- Spiritual well-being as a medical concept is promoted by these organisations, by eliciting the divine force in every individual which energises their physical, mental and psychological systems. Health care would be holistic only when spiritual dimension is added to these other concepts (Home Page | Sathya Sai International Organisation. http://www.sathyasai.org/. Accessed 24 July 2014)

11.2 The UNESCO Declaration on Bioethics and Human Rights

"Conscious of the unique capacity of human beings to reflect upon their own existence and on their environment, to perceive injustice, to avoid danger, to assume responsibility, to seek co-operation and to exhibit the moral sense that gives expression to ethical principles", UNESCO in its Declaration on Bioethics and Human Rights adopted by the General Assembly on 19 October 2005 included social responsibility and health as the central focus of Article 14 and appealed to all governments of the signatory countries to lay emphasis on promotion of health and social development for the people of the respective countries. The Article further states that since enjoyment of the highest attainable standard of health is one of the fundamental rights of every human being without distinction of race, religion, political belief and economic or social condition, progress in science and technology should advance access to quality health and essential medicine especially for the health of women and children, access to adequate nutrition and water, improvement of living conditions and environment, elimination of marginalisation of any group and reduction of poverty and literacy (UNESCO 2005).

Having seen what Hinduism prescribes as the *dharma* or responsibility of individuals and communities towards social welfare and the UNESCO's expectation from the member countries in achieving health through social responsibility by all sectors, we will now examine how the Indian government is addressing this issue since its independence in 1947 from the British empire.

11.3 The Constitution of India

The Constitution of India adopted by the Constitutional assembly on 26 November 1949 came into force on 26 January 1950 when India became a sovereign Republic nation. The desire for complete freedom and development with equity and justice, which the Indians were looking forward all through the period before the independence, got an opportunity for full expression in the form of the Constitution. It is unique both from the point of view of contents and size and form and is considered to be the longest and most detailed Constitution in the world.

Apart from outlining the Fundamental Rights, Directive Principles and Fundamental Duties of the citizens, the Constitution highlights the necessity of the integrations of social and economic rights with political freedom. Thus, the Indian Constitution eminently indicates the Indian state being a welfare state. Pandit Jawaharlal Nehru, the first Prime Minister of independant India said, "Political freedom is only a means to an end, the end being the raising of the people to higher levels and hence the general advancement of humanity", which is the basic philosophy of the Indian Constitution.

Although health as a separate sector does not appear in many places of the Constitution, there are indirect and tacit references to the health of the people. It is clear that the Constitution expresses its deep concern for the human, animal, plant and aquatic health in India. In the Indian federal system, different sectors of government have different responsibilities that concern various aspects of health. It also stipulates certain duties for the citizens towards contributing to the promotion of health in the country. The Indian Constitution has also exhibited its appreciation for the comprehensive perception of "Health for All". In addition to the government, many non-governmental and religious organisations are involved in health-care services for the community (Raghavendra Rao, and Panchamukhi. Health and the Indian Constitution, Centre for Multi-diciplinary Development Research Monograph Serial No. 7. http://cmdr.ac.in/editor_v51/assets/Mono-7.pdf).

11.4 Corporate Social Responsibility (CSR)

The World Business Council for Sustainable Development states that, "Corporate Social Responsibility is the continuing commitment by business entities to behave ethically and contribute to economic development while improving the quality of life of the workforce and their families as well as of the local community and society at large". With the proliferation of the private sector and accumulation of enormous fortunes by taking advantage of the government deregulation and economic liberalisation and globalisation, many private companies in India have started promoting CSR and private philanthropic activities. Alhough in a nascent stage, recently several corporate bodies have partnered with many civil society organisations in the process of implementing their CSR programmes. This became a prominent feature specially in the context of the Companies Bill, 2009, of the Government of India which suggests that the companies are expected to earmark 2% of the net profits towards CSR activities every year (Mathew 2011).

In recent years, numerous conceptual studies have been conducted on CSR from various perspectives including spiritual and religious angles. An interpretation of Bhagavad Gita related to the concept of *Dharma* and *Karma* has been brought out vividly by Muniappan and Satpathy as an inside-out approach to CSR, which is the development of individual leader's self-conscience of his *Dharma* (duty) and *Karma* (action) (Muniapan and Dass 2008). Based on *Bhagavad Gita's* insights, CSR ideally begins with ISR (individual social responsibility) including personal social responsibility and moves beyond CSR to GSR (global social responsibility). These three responsibilities are the essential elements of *Dharma* and *Karma* of corporate leaders in the role they play in corporations in ensuring transparency, good conduct and governance towards the ultimate aim of achieving CSR.

11.5 Conclusion

The UNESCO Declaration on Bioethics and Human Rights addresses ethical issues related to medicine, life sciences and associated technologies as applied to human beings considering their social, legal and environmental dimensions and provides guidance to states or member countries for decision or practices of individual, groups, communities, institutions and corporations, public and private. This concept as applicable to human rights, fundamental freedom and human dignity has been examined in this paper through the Hindu philosophical approach existing from time immemorial towards duties and responsibilities of individuals towards self and society at large and the current guiding principles of the Constitution of India for protection of the fundamental rights of its citizens through directive principles and fundamental duties for establishing a welfare state. The role of religious institutions and modern corporate houses in promoting Health for All, abiding by the Hindu tenets of *Dharma* and *Karma* to execute their social responsibility, has also been elaborated.

References

"A Timeless Woman with a Timely Message." *The Interfaith observer*. http://www.theinterfaithobserver.org/journal-articles/2011/12/5/a-timeless-woman-with-a-timely-message.html. Accessed 24 July 2014.

Bhargava, Anju. The importance of seva and social justice for inner transformation," in https://www.huffingtonpost.com/anju-bhargava/hindu-aeva-and-social-jus_b_824360.html.

Bose, Manilal. 1998. *Social and cultural history of ancient India*, 68–79. New Delhi: Concept Publishing Company.

"Duties and Responsibilities, the Hindu Social Framework,". http://artstudio.co.za/Duties_and_Responsibilities_The_Hindu_Social_Framework.html.

Easwaran, Eknath. 2010. *The Bhagavad Gita*, 54. New Dehli: Jaico Publications.

Gandhi, M.K. 1958. *Hindu Dharma*. Ahmedabad: Navjivan Publishing House.

Hindupaedia. The Hindu Encyclopaedia. http://hindupedia.com/en/Main_Page. Accessed 5 June 2014.

"Home Page | Sathya Sai International Organisation.". http://www.sathyasai.org/. Accessed 24 July 2014.

Mahabharatha, Karna Parva (69.58).

Mahabharatha, Shanti Parva (109-9-11).

Mathew, Joe C. February 19, 2011. Companies bill: Firms keep fingers crossed on CSR. *Business Standard India*. http://www.business-standard.com/article/companies/companies-bill-firms-keep-fingers-crossed-on-csr-111021900066_1.html.

Muniapan, Balakrishnan, and Mohan Dass. 2008. Corporate Social Responsibility: A Philosophical Approach from an Ancient Indian Perspective. *International Journal of Indian Culture and Business Management* 1 (4): 408–420. https://doi.org/10.1504/IJICBM.2008.018622.

Muniapan, Balakrishnan, and Biswajit Satpathy. 2013. The 'Dharma' and 'Karma' of CSR from the Bhagavad-Gita. *Journal of Human Values* 19 (2): 173–187. https://doi.org/10.1177/0971685813492265.

Parashara Smriti, Madhvacharya.

Pattanaik, Devdutt. 2010. *Jaya: An Illustrated Retelling of the Mahabharata*. Gurgaon: Penguin Global.

Purva Mimamsa, Rishi Jamini (1–2).

Raghavendra Rao, K., and P.R. Panchamukhi. *Health and the Indian Constitution*, Centre for Multi-diciplinary Development Research Monograph Serial No. 7. http://cmdr.ac.in/editor_v51/assets/Mono-7.pdf.

Sharma, Arvind. 2002. The Hindu Tradition. In *Religious traditions & healthcare decisions: Religious beliefs and their application in health care: A reference guide*. Chicago:. Park Ridge

Centre for the study of Health, Faith and Ethics. https://www.ethicshare.org/node/719198. Accessed 11 Dec 2017.

Swami Nikhilananda. *Hindu ethics*. New York: Ramakrishna Vivekananda Centre. http://www.hinduism.co.za/ethics.htm. Accessed 15 June 2014.

Swami Ranganathananda. 1971. *Introduction to the message of Upanishads*. 2nd ed. Mumbai: Bharatiya Vidya Bhawan. http://www.vedanta.gr/wp-content/uploads/2012/03/SwRang_MsgUpa_Intro_ENA5.pdf.

Swami Sivananda. *All about Hinduism*. A Divine Life Society Publication, 1999. http://www.dlshq.org/download/hinduismbk.htm.

UNESCO. *UNESCO* Declaration on Bioethics and Human Rights, 2005. http://www.unesco.org/new/en/social-and-human-sciences/themes/bioethics/bioethics-and-human-rights/.

"VHP's Social service activities," *The Hindu*. (18/12/2011). http://www.thehindu.com/todays-paper/tp-national/tp-kerala/vhps-social-service-activities/article2725683.ece.

Viswanathan, Kalyan. 2008. Bhagvad Gita and Mahatma Gandhi – Dharma and ahimsa. In http://blogs.ivarta.com/Mahatma-Gandhi-Bhagvad-Gita/blog-106.htm.

Wangu, Madhu Bazaz. 2009. *Hinduism*, 4th ed, ed. Joanne O'Brien and Martin Palmer. New York: Chelsea House Publications.

Whaling, Frank. 2013. *Understanding the Brahma Kumaris*, 65. New Delhi: Pentagon Press.

Chapter 12
Karma in Bioethics Reasoning: A Hindu Response to Vasantha Muthuswamy

John Lunstroth

Tat tvam asi (that thou art)

CHANDOGYA UPANISHAD

Abstract Some bioethicists explore the question of whether there are unique ways of understanding bioethics using philosophical concepts from the Indian philosophical tradition. One of the best known of such concepts is *karma*. Karma is a Sanskrit word that in common parlance means "as you sow, so shall you reap." But, like most philosophical concepts, for philosophers it is an immensely complicated and nuanced concept that is used in somewhat different ways in the diverse complex philosophical systems that constitute the Indian philosophical tradition. I argue it is inapt for bioethical reasoning. To the extent bioethics is practical ethics, then the concept is inapt because its native use is to understand certain experiences that arise in the spiritual life, especially in meditation (*sadhana*), the common practice among Indian philosophical traditions. To the extent bioethics is theoretical ethics, the matters touched on in karmic analyses are far removed from sociopolitical institutions within and from which bioethics derives its meaning. Bioethics is a school of ethics. Ethics is meant to guide action in the political sphere. The concept of karma categorically leads to political quietism or inaction. Even if one seeks to act thinking that "good" karma will arise from one's actions, the concept of karma philosophically rejects the idea that there is any ultimate "good" karma, and moreover, one can never know whether one is fulfilling existing karma or creating new karma when one acts. Therefore, the use of karma in bioethics discourse is not recommended.

Keywords Hinduism · Hindu bioethics · Karma · Practical ethics · Theoretical ethics

J. Lunstroth (✉)
Health Law & Policy Institute, University of Houston Law Center, Houston, TX, USA

J. Tham et al. (eds.), *Religious Perspectives on Social Responsibility in Health*, Advancing Global Bioethics 9, https://doi.org/10.1007/978-3-319-71849-1_12

I am going to compare two living religious tradition, Christianity and Hinduism. On the one hand, Christianity recognizes itself as a religion. It has at least three large institutional bodies, Protestantism, Orthodoxy, and Catholicism, one of which, Catholicism, is also a political state recognized as such by the United Nations and other states. On the other I will be looking at Hinduism. The label of Hinduism as a reference to a religion is used by the West almost exclusively, and by some citizens of India, to refer to a hypothesized and often politicized non-Abrahamic religion of the peoples living in the Indian subcontinent. That label is less than 200 years old. The Westerner looks at religion in the West, and when it looks at India, it looks for the same thing, and it thinks it sees it. But suffice it to say that what unites people in the Indian subcontinent is a way of life that is, generally speaking, deeply spiritual, all-inclusive and exceedingly tolerant, a mystical practice, and not a specific set of beliefs. It embraces Islam, Christianity, the plethora of indigenous and imported gods and goddesses, Buddhism and its various deities and systems of knowledge, and the most sublime systematic philosophies of the various peoples of the subcontinent. Instead of the word "Hindu" and its derivatives, I am going to use the word "Indian."

In the following comments, the reader is going to hear echoes of Alan Watts and his successors who in the West found the East and easily saw the great and enduring similarities between the two. In the process of describing the similarities though, they often overlooked not only pesky details but profound philosophical differences. I begin then alluding to what I think is the main philosophical or theological difference and expressing respect for both understandings. I will then look at one of the common goals of both, social justice, and use that as a focal point to reach into each tradition to find commonalities. It is in this that I try to achieve the goal of the UNESCO Chair in Bioethics and Human Rights, to seek convergence among the various traditions.

Christianity is monotheistic, and the Indian tradition is not. Islam, on a historicist view, had complex encounters with both and has elements of both, so in the interests of brevity and coherence, I only touch on it lightly. The monotheism of Christianity is of a specific form organized by St. Augustine. I argue elsewhere that Christianity prior to his innovations was probably more like Hinduism in a particular and relevant way.[1] Augustine theorized a unified, singular historical narrative for all of creation and all time, over which the Christian God, in unity with Christ and the Holy Spirit, ruled. Later amendments to the theory by the Universalist Popes of the Late Middle Ages clarified that God was the absolute and universal source of all law within this single historical narrative. It is this single and unified narrative that, in my opinion, set the stage for the development of the unique secular approach to science that was institutionalized in the Enlightenment and which has given rise to the modern oligarchic global economic culture. Augustine materialized a Christian vision of a Platonic political and spiritual body. There is one God and only one way to Him, through Christ, his Son, and Identity.

[1] John Lunstroth, History, Universalism and Religion: West and East (2015), at https://unescobiochair.wordpress.com/2015/01/29/history-universalism-and-religion-west-and-east/

This is contrasted with the various ways of life of the Indian subcontinent, in which God was early recognized as being essentially formless, which means any form that could lead to the presence of God was valued. In this sense the Indian tradition could be said to generalize the Christian religious form. India and Christianity both recognize God, and this is surely the same God. But whereas Christianity holds there is only one Jesus Christ, the Indian tradition recognizes many Christlike figures, including both Buddha and Christ. The Indian does not find Christianity different at all, except for the way the injunction that "there is only one Way to God" is interpreted. To the Indian, that means the way to God is only through the Teacher, and who one's Teacher is can be different from person to person and family member to family member, whereas the Christian holds there is only one Teacher. Philosophically God is formless (or not), and therefore the notion of what is truly universal law remains much more abstract. The Indian notion of the universal is Aristotelian, not Augustinian; history proceeds and is manifested in multiple overlapping and intersecting unities, not within one narrative with one set of laws. Each great Teacher becomes a lawgiver in the unity with his or her disciples. For any seeker of God, the only way to get there, in general, is through the seeker's Teacher.

I leave the differences here. Both traditions value helping and serving the needy, and both have conferred measureless actual and material benefits to humans, and sometimes nonhumans, in the name of God. I will understand this as a shared value and briefly explore the notions of duty, *dharma*, and *karma* as discussed in the papers of Muthuswamy[2] and de los Rio[3] on the Indian tradition and Christianity, respectively.

Both the Christian and the Indian are inculcated in a culture of social justice. There are many kind souls in each tradition that serve the needy out of the depths of their saintly natures. For many of them, the reference to the laws of their religious or spiritual affiliation is secondary to their sense of duty to those they serve. They have internalized their connection to God and their Teacher. Of course, there are others who serve for more worldly reasons and who aspire to be like their more saintly brothers and sisters.

Rios argues that through Love the Christian is able to see Christ in everyone. That Godly Love drives the Christian to serve the other, and in the process of serving the other, "God's forgiveness … liberates the human being of its guilt allowing always for a new beginning."[4] The Christian is forgiven and renewed in Godly service and reconciled with God. The act Loving and serving God in the other frees the Christian from historical necessity.

Muthuswamy focuses on the extensive literature in the *Sanskrit* (i.e., Indian) tradition about the importance of service to the needy in the doctrines of *dharma* and *karma*. *Dharma* refers to the external order of the universe and the need of the

---

[2] Vasantha Muthuswamy, Hinduism and Social Responsibility (2016).

[3] Elizabeth de los Rios Uriarte, Christian Response on the Hindu Perspective on Social Responsibility by Vasantha Muthuswamy (2016).

[4] Rios, Part 3.

individual to act within that natural order. *Karma* refers to a different kind of natural system that concerns the individual (and other entities, but I will not discuss them). An individual in this system is understood as a soul embodied in various kinds of increasingly gross matter. For simplicity's sake, these can be understood to be a mental body, an emotional body, and a physical body. At the death of the physical body, the emotional and mental bodies live on; and on the death of the emotional body, the mental body lives on, until it dies; and then the soul shines unhindered. That individual soul is to God as Christ is to God. *Karma* refers to a kind of natural law that comes into effect when the soul enters into the bodies. The soul becomes enamored of its experiences in the bodies, and it has thoughts, feelings, and acts infused with desire for pleasure or other attachments. Mental, emotional, and physical acts undertaken with desire for an outcome become part of an illusion that seizes the embodied soul (e.g., *vasana*), and in this state, those acts create conditions that get repeated endlessly in the wheel of *samsara* or illusion, dragging the soul with them. There are different kinds of *karmas*, but the various distinctions are not important to this paper. This anthropology explains reincarnation as well. When the embodied and ensnared soul is compelled by desires and habits (*karmas*), it in turn becomes embodied in a physical form, albeit having forgotten its origins and true nature. The mental body lives longer than the emotional body, and the emotional body longer than the physical, so when the more subtle bodies come into the physical, they are continuous with earlier experiences in a physical body. The foregoing is a kind of theosophical description meant for Western sensibilities and is not meant to refer to any particular philosophical school.

Muthuswamy explains clearly how this cosmic drama is maintained by the *dharma* of social justice and compassion for the other. *Dharma* refers, both descriptively and normatively, to the overarching connections between all the souls and their various embodiments. When love for the true self of the other is the dominant sociopolitical value, then love of God and truth prevail in the sociopolitical order. When the individual embodied soul sees God in the other and serves the other without desire for personal gain, then those acts lead to liberation from the wheel of birth, death, and rebirth. The technical descriptions of how it leads to liberation are beyond the scope of this paper, but the kind of summary offered by Rios to describe how it leads to liberation for the Christian can be applied here. In this regard, one significant difference between Christianity and life in India is that the Christian would understand the Indian lifestyle as being an entire culture of mystical engagement with God.

In the following brief excerpt from the *Bhagavad Gita*, Chap. 2, here translated by Swami Prabhupada, one can see a textual basis for what I have said above and what Muthuswamy says. Arjuna, the disciple, is being addressed by Krishna, the Lord:

44. Of [those who are] attached to enjoyment and lordship [and have] "carried-away" minds – [their] wisdom-faculty, [which is of] the essence of determination, is not settled in ecstasy.

45. The triad of primary qualities [of the manifested universe] is the subject-matter of the Vedas. Become free of the triple primary-qualities, free of the pairs-of-opposites, and, O Arjuna, abide always in *sattva*, without [trying to] gain or keep [anything]. [Be] Self-possessed!
46. As much use [as is] a water-reservoir flooded with water all-round, so much [use is there] in all the Vedas for the knowing Brahmin.
47. In action alone is your rightful-interest (*adhikāra*), never in [its] fruit. Let not your motive be the fruit of action; nor let your attachment be to inaction (*akarman*).
48. Steadfast in Yoga, perform actions abandoning attachment, O Dhanamjaya, [always] remaining the same in success and failure. Yoga is called equanimity.
49. Far inferior indeed is [mere] action than *buddhi-yoga*, O Dhanamjaya. Seek refuge in the wisdom-faculty! Pitiful are those whose motive is the fruit [of action].
50. The *buddhi*-yoked leaves behind here [in this world] both well-done and ill-done [actions]. Hence yoke yourself to Yoga. Yoga is skill in [the performance of] actions.
51. The wise [who are] *buddhi* yoked, who have renounced action-born fruit, who are liberated from the bondage of birth [and death] – they go to the region [that is] free from ill. …
70. Just as the waters enter the ocean, full [yet having] unmoving ground, so all desires enter him; he attains peace [but] but not the desirer of desires.
71. That man (*pumān*) who, forsaking all desires, moves about devoid of longing, devoid of [the thought of] "mine," without ego-sense – he approaches peace.
72. That is the brahmic state, O son-of-Pritha. Attaining this, [a person] is no [longer] deluded. Abiding therein also at the end-time [i.e., at death], he attains extinction in the world-ground (*brahma-nirvāna*).

The text refers to yoga (a technical word) as the means by which to fulfill one's *dharma* and escape the bonds of karma. Elsewhere in the text, the nature and role of the Teacher in this process are made clear.

Muthuswamy's paper focuses on the textual and doctrinal laws of *dharma* and *karma* as providing what a Westerner would call an ethical or political justification for serving the needy. Rios, in her paper, focuses on Love as the transcendent experience that comes from selflessly serving the needy and the resultant liberation. I think Muthuswamy's purpose was different from Rios', as had she turned her focus slightly she would have salted her paper with the perspectives I add. The Sanskrit tradition is at least 2500 years old, and the teachings on what a Westerner would understand as sociopolitical and natural order, and of the golden rule, are vast and deep, just as they are in the West. But in the Indian tradition, they are almost uniformly grounded in mysticism, whereas the West long ago rejected the mystical perspective.

There are grounds to understand a profound convergence between Christianity and the Indian tradition in the impulses and injunctions to serve the needy in the Name of God; to see God in the other, especially in the suffering of the other; and to find one's way from separation and bad habits into the presence of God or liberation. There is an immediate and direct identity between the two traditions in the centrality of Godly Love as the experience that disconnects one from the necessities of history and of the necessity for a Teacher to show the way to the liberating experience of such Love.

Chapter 13
Reflections on Destiny, Forgiveness, and Reconciliation in Social Responsibility: A Christian Response to Vasantha Muthuswamy

María Elizabeth de los Ríos

Abstract In Hinduism there is an internal law that moves and guides all actions of a person which is "Dharma," which literally means "to keep" or "maintain." This law is sustained in itself by the practice of four virtues: love, truth, non-violence, and ethics. On the contrary, from a Christian perspective, the fundaments of social responsibility have the living image of a God converted into man through the figure of Christ Resurrected. The fundaments, the logic, and the consequences of social responsibility in both religions are different; nonetheless, in Hinduism the human being seems not to be able to escape a destiny established beforehand, while in Christianity there is always a possibility of a new start and a new opportunity to recognize God among us and, in such a manner, to be able to walk with Him; both religions wager on doing good and mutual assistance.

Keywords Hinduism · Christianity · Social responsibility · Dharma · God

13.1 Introduction

When Mother Teresa of Calcutta received her "calling" to establish the Congregation of the Missionaries of Charity, it happened through an encounter with a vagabond in the train station of Darjeeling who barely uttered the words, "I thirst," before this nun of the Sisters of Loreto. There and then, Agnes Gonxha Bojaxhiu, Mother Teresa, received in her innermost being the impact that would take her to serve the poorest of the poor and establish a new religious order that would pursue such a goal.

M. E. de los Ríos (✉)
Faculty of Bioethics, Anahuac University, Mexico City, Mexico

J. Tham et al. (eds.), *Religious Perspectives on Social Responsibility in Health*, Advancing Global Bioethics 9, https://doi.org/10.1007/978-3-319-71849-1_13

I mention this story to illustrate two important elements I will elaborate upon reading the article by Muthuswamy: first, what constitutes the basis of mutual help in the conception of social responsibility in Hinduism and, second, the internal logic of such basis. Finally, I would like to review some consequences which (fully aware that I could be wrong as I'm not a practitioner of Hinduism) may arise from the internal dynamics of the Hindu way of life regarding social responsibility and contrast them with the Christian faith so as to foster further dialogues and discussions.

My presentation will therefore consist in the analysis of the foundations of social responsibility in Hinduism as viewed from a Christian perspective, the treatment of the internal logic pursued by such foundations, and the consequences that I believe are derived from the latter.

13.2 Foundations, Logic, and Consequences of Social Responsibility in Hinduism

As Muthuswamy states, in Hinduism there is an internal law that moves and guides all actions of a person – *Dharma*, which literally means to "keep" or "maintain." This law is sustained by the practice of four virtues: love, truth, non-violence, and ethics. To orient oneself in the search and keeping of these virtues, a Hindu believer must above all pursue his spiritual perfection and then consequently does good to the others. Thus, *Dharma* is a consequence of the mandate of personal rights and duties to carry out the social good so that cosmic order may ensue.

This cosmovision has a special relevance to those who practice Hinduism as stated in the article in question, since there is a connection not only between individuals and the society in which they live but also with the cosmos. Hinduism believes that every species and every living being maintain a specific place by keeping their respective duties. When they comply with the *Dharma* of their appropriate place, harmony results. In this manner, he who carries out his duties makes others comply their own duties, and when everyone complies, the universe is in order.

We may further inquire about the nature of this compliance with specific duties, since the origin of such obligations is not very clear. Muthuswamy explains that in Hinduism, birth already represents a debt. When one comes into this world, a debt is incurred that has at least four aspects: debt toward the gods, debt toward one's ancestors, debt toward the wisdom of life, and debt toward the universe. The last debt is with the rest of other beings (animals, plants, etc.) This indebtedness explains why everyone must carry out their respective tasks.

These ideas together with the prevailing belief in reincarnation explain the eternal enchainment to debts that are paid through consecutive births but without the possibility of ever repaying them all in full. This in turn brings us to the idea of an inevitable fatalistic destiny or *karma* that pervades all lives. There is a tense and troubling sense that human life is condemned to an interminable cycle of reincarnations to pay for the debts incurred – but since each birth is a debt, absolute liberation is unattainable.

Likewise, since all actions represent an additional link in the chain of fatalistic *karma* of all living beings, this "punishment" or "sentence" becomes the debt of each preceding life. In this manner, each action – even those in themselves directed toward the good or benefit of others – has as its primary goal the need to pay for a fault committed in order to escape future punishment of an undesired reincarnation. Charitable actions are not carried out for their own sake, and less for the sake of others, but out of the fear of eternal punishments.

If I am correct, the source from which social responsibility emanates in Hinduism comes from this conception of an acquired debt and its consequential payment out of fear of punishment. Responsibility lies first with myself and then with others only when they interfere in some manner with my own destiny.

13.3 Logical Foundations and Consequences of Social Responsibility in Christianity

On the contrary, the foundation of social responsibility for Christianity is – as in the case of Mother Teresa – the living image of God made man through the figure of the Resurrected Christ.

Christianity is based on the belief in an incarnate God who lives among us. As the Gospel states, "As you did it to one of the least of these my brothers, you did it to me" (Mt. 25:40). Help offered toward others is based above all on the deep and profound preoccupation for the instauration of the Kingdom of God on Earth, as the *Compendium of the Social Doctrine of the Church* (2005) mentions:

> Human activity in history is of itself significant and effective for the definitive establishment of the Kingdom... Such activity, when it respects the objective order of temporal reality and is enlightened by truth and love, becomes an instrument for making justice and peace ever more fully and integrally present, and anticipates in our own day the promised Kingdom. (n. 58)

Pope Leon XIII emphasized the "social question" in his Encyclical *Rerum Novarum* which Pope Paul VI follows in *Populorum Progressio*. They represent a living example of the attention Christ paid to the neediest in his public life. As Paul VI says in *Populorum Progressio* (1967):

> In the present day, however, individual and group effort within these countries is no longer enough. The world situation requires the concerted effort of everyone, a thorough examination of every facet of the problem – social, economic, cultural and spiritual.

> The Church, which has long experience in human affairs and has no desire to be involved in the political activities of any nation, 'seeks but one goal: to carry forward the work of Christ under the lead of the befriending Spirit. And Christ entered this world to give witness to the truth; to save, not to judge; to serve, not to be served.' (n. 13)

Jesus had predilection for the poor, widows, strangers, the sick, and the marginalized. Those who "live on the fringes," in terms used by Pope Francis, are cared for and should be a concern of Christians in both public and private institutions. This disposition exemplified an internal logic which is one of Love.

These actions directed toward the good of oneself and others are based on the solid conviction that Christ is "The Way, the Truth and the Life" (Jn.14:6) and that he lives among us and is one with us. When Mother Teresa hears the words "I thirst," she didn't hear a poor man lying among the railroad tracks. She heard none other than Jesus who asked her to give him something to drink, like the Samaritan woman at the well (Jn. 4:5, 42), or his last words on the Cross exclaiming that he was thirsty (Jn.19:28). He was present there in that man – neglected, sick, wounded, hungry, and thirsty – and she understood that Christ was alive and Resurrection acquired its truest and most profound meaning.

It is not so much as "having" to do something in Hinduism. In Christianity, we "want" to do it because we hope and we live in the Resurrected Christ. Actions are not taken out of "fear" of punishment, since for Christians, an eternal life begins here and now, in the present moment where eternity becomes real. Christian beliefs are centered on the glory and the joy of finding Christ in our lives and not on punishment or penalties. Thus, joy should be the most significant characteristic of Christian living.

Christians do not believe in a fatalistic destiny or an incalculable chain of debt that must be paid since Someone has already paid for them in our stead – Someone who came, suffered, and died for us so that we can have "life in abundance" (Jn.10:10).

The prophet Isaiah in the Song of the Suffering Servant speaks of a just man unjustly executed. The logic that moves the Christians is not one of punishment or fear, but the logic of love. It also implies an apparently contradictory, inverse logic – that of forgiveness and reconciliation (Cortés 2011). In contrast to a *karma* that affirms that all actions – good or bad – come back to us in a circular and infinite manner, Christianity proposes to break this cycle by inserting forgiveness and reconciliation.

Forgiveness as a theological condition allows us to vanquish the initial offense by giving it to the other as an offering of love. That was what Jesus did on the Cross. He redeemed the human condition by assuming it in all its intensity. This offers a new possibility that goes beyond human reason and represents the novelty of love and charity. As theologian Paul Tillich (1973) mentions:

> Forgiveness means reconciliation even though hostility. God's forgiveness is unconditional. There is no sign, not even the smallest, in men that makes him subject of forgiveness. As a gift, it is too big. God's forgiveness enters in our ambiguities but does not asks for us to stop being that ambiguous.

Reconciliation and forgiveness of other's offenses, and above all God's forgiveness, liberates us of our guilt and leads always to a new beginning. As Saint Paul says, "it is no longer I who live, but Christ who lives in me" (Gal. 2: 20). Christians focus not so much on the lack of condemnation for this would be a negative or excluding logic. Instead they offer the maximum act of love: to give their lives for another knowing that these "others" are "other Christs."

13.4 Conclusion

Dharma according to Hinduism tries to recover the lost order of things from past lives at the moment of birth. Human actions are directed toward doing good more as obedience to the internal order of things than a transcendent recognition of the needs of others. *Dharma* looks for order and obtains it through *karma* which sets each one in his appropriate place. According to this logic, there is little space for liberty or choice since one's fate is set and chance has already been cast.

There is a different dynamic of love in Christianity. It is strange and disruptive as it inserts the possibility of recognizing the other as really Christ incarnated in the poorest and most vulnerable. The desire to turn toward Him and live in Him is the basis for reaching out to the others. This approximation extends to the most "distant," to those who have offended us, even most seriously. Through forgiveness and reconciliation, they are granted liberty they have lost through sin. In the struggle for justice and peace, they are once again our "brothers" or, as Saint Ignatius of Loyola used to sign his letters, "as friends in the Lord."

The basis, logic, and consequences of social responsibility in both religions are therefore quite different. Both religions wager on doing good and mutual assistance for "the highest standards of health, without distinction of race, religion, or political beliefs" as Article 14 of the UNESCO Declaration on Bioethics and Human Rights states. Both promote solidarity and responsible actions toward others. The vision that underlies each one is what makes them differ. Nonetheless, a latent and current truth in the saying of Saint John of the Cross applies to them both: "In the sunset of our lives we will be judged by love."

References

Compendium of the Social Doctrine of the Church. 2005. *Pontifical council for justice and peace.* Vatican: Libreria Editrice Vaticana.

Cortés Fabela, Maria De los Ángeles Amelia. 2011. Un judío hereje. In *Acercamientos diversos a Jesús de Nazareth,* ed. Sánchez Zariñana, Humberto José. México: Universidad Iberoamericana.

Tillich, Paul. 1973. El nuevo ser. Barcelona.

Part V
Islam

Chapter 14
Social Responsibility and the Moral Obligation Toward Providing Healthcare: An Islamic Ethico-legal Analysis

Aasim I. Padela

Abstract In this essay, I present an Islamic ethico-legal perspective on the relationships between social responsibility and the state's provision of healthcare as part of a continuing dialogue between academics, policy stakeholders, and representatives of religious communities about the UNESCO Declaration on Bioethics and Human Rights (UNESCO 2006). I briefly recount the notion of human rights in an Islamic moral universe and then proceed to describing conceptual analogues within the Islamic tradition for the principles of social responsibility and community obligation. I use these to construct an argument for the state's obligation to provide a certain type of healthcare to its citizenry. I next relate two real-world examples of Muslim stakeholders operationalizing the Islamic ethos of community obligation regarding health through servicing the religious needs of Muslim patients. By incorporating both theological and applied ethics in this commentary, I hope to bring into view the concordances and discordances between the Islamic ethico-legal tradition (in theory and practice) and the theoretical underpinnings of human rights doctrine and its related bioethics discourses.

Keywords Islamic bioethics · Right to health · Community health · Social responsibility · Health systems design

A. I. Padela (✉)
Initiative on Islam and Medicine; Faculty Maclean Center for Clinical Medical Ethics, University of Chicago, Chicago, IL, USA
e-mail: apadela@uchicago.edu

J. Tham et al. (eds.), *Religious Perspectives on Social Responsibility in Health*, Advancing Global Bioethics 9, https://doi.org/10.1007/978-3-319-71849-1_14

14.1 Introduction

In the present era of global interconnectedness, transnational institutions such as the United Nations often adopt human rights arguments in order to inspire a global moral consensus on various social policies. While this approach may avoid partisan argumentation based on religious principles and cultural values that are not universally accepted, human rights doctrines have their own sets of assumptions regarding human needs and societal end goals that may undermine their own universality. The historical example of Muslim nations and stakeholders fashioning their own *Declaration of Human Rights in Islam* and the *Universal Islamic Declaration of Human Rights* in response to the United Nations' *Universal Declaration of Human Rights* (UDHR) illustrates that, at least for some Muslim stakeholders, human rights lexicon is not wholly effectual for rooting moral claims (Cairo Declaration on Human Rights in Islam 1990; Universal Islamic Declaration of Human Rights 1981; The Universal Declaration of Human Rights 1948). Indeed these responsa replace the grounding of fundamental moral obligations in the shared human nature of individuals with references to Islamic sacred law, the *Sharī'a,* as the origin of moral obligations. This seemingly small change in how one grounds rights and obligations is profoundly important to dialogue over the applicability of the United Nations Educational, Scientific and Cultural Organization's (UNESCO) Declaration on Bioethics and Human Rights to Muslim societies and Islamic contexts.

In this essay, I offer an Islamic ethico-legal perspective on the relationships between social responsibility and the state's provision of healthcare as part of a continuing dialogue between academics, policy stakeholders, and representatives of religious communities about the UNESCO Declaration on Bioethics and Human Rights. Since this paper marks the UNESCO-sponsored Fourth International Bioethics, Multiculturalism and Religion Workshop where Article 14 of the UNESCO Declaration (reproduced in the introduction to the volume) was deliberated over, I will begin by briefly recounting the notion of human rights in an Islamic moral universe and then proceed to describing social responsibilities and the state's responsibility for the healthcare of its citizens with reference to several Islamic ethico-legal concepts (Padela 2017). At the end of the paper, I will recount two real-world examples of Muslim stakeholders operationalizing their Islamic values by attending to the religious needs of Muslim patients. By incorporating both theological and applied ethics in this commentary, I intend to bring into view the concordances and discordances between the Islamic ethico-legal tradition (in theory and practice) and the theoretical underpinnings of human rights doctrine and its related bioethics discourses.

14.2 Locating Notions of Social Responsibility and Human Rights Within the Islamic Tradition

Social responsibility as a normative ethical principle is used "as the way of defining the moral duties" that entities (corporations, governments, and/or individuals) have toward the benefit of the larger society (International Bioethics Committee 2010). This principle is deemed by UNESCO to be a fundamental one because it is grounded within "social relationships" and is "part of what has traditionally been called moral obligations" that "cannot be imposed by others or by the State" (International Bioethics Committee 2010). In this way, social responsibility resides within a human rights framework in which one can make claims upon state authorities "simply because one is a human being" (Moosa 2004). Just as belonging to the human race (having a shared biology) creates the condition for arriving at fundamental human rights, so being part of a society—having "social relationships" as it were—creates the conditions for moral obligations based on social responsibility. Accordingly, the Declaration's Article 14 grounds the state's moral obligation to provide for the healthcare of its citizens in social responsibility.

A human rights-based moral framework appears to be insufficient for grounding moral obligations in Islam. As Sachedina notes "there is no unequivocal language of human rights per se in Islamic sources" (Sachedina 2009) because according to Islamic moral theology (*uṣūl al-fiqh*), rights do not inhere within humankind and do not emerge organically from social relationships; rather obligations ensue from the Divine ascription of a moral status to human actions: "God is the one who confers rights on persons via revealed authority" (Moosa 2004). This attribution is gleaned from a close reading of the Qur'an and Prophetic Sunnah—the two parts of the Islamic revelation. Notions of social responsibility, therefore, are insufficient by themselves to ground an Islamic obligation toward healthcare. Instead, social obligations require substantiation within the sacred source texts of Islam and must be reflected into Islamic law in order for these notions to ground moral duties.

To embark on that sort of project requires bearing in mind two constructs within Islamic law: the *ḥuqūq al-ibad/ḥuqūq Allah* heuristic and the concept of *farḍ al-kifāyah*. As will be described below, *ḥuqūq al-'ibād* and *farḍ al-kifāyah* are partial analogues to the concepts of social responsibility and social obligation and can provide the foundational building blocks for the religio-cultural translation of Article 14's posited obligations into Islamic ethico-legal lexicon.

Within the Islamic ethico-legal tradition, there are two primary kinds of "rights" (*ḥuqūq*): the "rights" of God and the "rights" of humankind. The term for a human being's obligations to other human beings is *ḥuqūq al-ibad*. *Ḥuqūq Allah* refers to the "rights of God," yet since theologically God is free from any need, such a translation is somewhat problematic. These "rights" are not freedoms that anyone else has granted to God, but neither are the "rights of God" simply obligations human beings owe to God. Rather *ḥuqūq Allah* is a term of art used by Islamic jurists to classify two types of obligations. The first are duties of a devotional kind (e.g., the five pillars of worship) that have religious ends and are thus owed to God, and the second are obligations that relate to social goods that are "beneficial to the community at-large" (Sachedina 2009; Moosa 2004). To further clarify this second type of obligation, Islamic jurists consider the well-being of society as a "right" owed to God that a Muslim ruler must uphold in "light of his imperium over society" (Emon 2006).

By contrast, the "rights of humankind," *ḥuqūq al-ibad*, involve private interests and reside at the individual level, including, for example, the "right to own the object (one) has purchased" (Kamali 2003). A third category of "rights" that are shared by God and humankind also exists encompassing activities in which both public and private interests are at stake. Islamic jurists use the rights of God/rights of humankind rubric to identify, and balance between, competing individual and societal interests. As will be illustrated below, plausible arguments can be made to classify the provision of healthcare as part of *ḥuqūq Allah* as well as part of *ḥuqūq al-ibad*.

Farḍ al-kifāyah refers to communal obligations in Islamic law (Fard al-Kifayah 2003). This category of actions is the responsibility of the community as a whole, and if discharged by a sufficient number of individuals, the remaining Muslim members of the community are freed from the responsibility before God. Conversely, if the "communal obligation is not sufficiently discharged, then every individual Muslim must act to address the deficiency," and each member carries sin and may incur afterlife penalty until the obligation is discharged (Fard al-Kifayah 2003).

14.3 The Preservation of Health As a Higher Objective of Islamic Law and the Provision of Healthcare As Part of *Ḥuqūq Allah*

With respect to social responsibility and the obligation to provide healthcare, Muslims must consider whether the provision of healthcare is part of *ḥuqūq al-'ibād* or *ḥuqūq Allah* and whether it is a *farḍ al-kifāyah*. With respect to the former question, one could certainly assert that the provision of healthcare is beneficial to the community at large and secures more than just private, individual-level interests when such provision is nondiscriminatory, i.e., the healthcare system provides unfettered access to a category of healthcare services for all inhabitants under the governance of a state authority. This line of argumentation suggests that when healthcare is not provided to individuals living under a Muslim state, public interests are at risk and a right of God is not being respected.

However, before elevating a specific public interest into the category of "rights" of God that a Muslim state authority is charged to protect, scriptural evidences—Qur'anic verses or Prophetic statements—are needed to support the argument that a particular social good or public interest is one that God through sacred law aims to preserve. While juridical reasoning can be employed to incorporate social goods and public interests not specifically mentioned by scripture within *ḥuqūq Allah* (Emon 2006), the closer the linkage is between a specific public interest and a scriptural source text, the stronger the evidence is for a definitive moral obligation. Islamic law has a classification schema for what constitutes a public interest—*maṣlaḥah*—and has guidelines under which *maṣlaḥah* can be used to generate moral law. In addition to this specific meaning of a defined, concrete public interest, *maṣlaḥah* also refers to the general practice of using public interest-based arguments within Islamic ethico-legal deliberation. *Maṣāliḥ* (plural form of *maṣlaḥah;* interests) can be private (*khāṣṣah*), general (*'āmmah*), or have both dimensions at the same time.

The eleventh–twelfth-century Islamic polymath jurist-theologian al-Ghazali defines *maṣlaḥah* as a combination of accruing benefit and avoiding harm and "explains 'what we mean by interests are those interests that conform specifically to the objectives (*maqṣūd*) of Islamic law, which are five: the preservation of religion, life, the intellect, lineage, and property. Anything which safeguards and guarantees those five principles is an interest. And anything that is contrary to realizing these five is a harm and corruption, the removal of which is also an interest'" (Abdur-Rashid et al. 2013). The aforementioned five aims of Islamic law are collectively referred to the higher objectives of Islamic law, *maqāṣid al-shar'iah.* While the process and scope are debated, the four Sunni schools of Islam appear to allow for *maṣlaḥah* to serve as grounds for a moral obligation when it comes to protecting the essential (*darūrī*) aspects of the five higher objectives of Islamic law (Kamali 2003; Yacoub 2001).

Notably, the usage of *maṣlaḥah* is bounded in that the public interest identified, the specific *maṣlaḥah*, must be definitive and not speculative, it must not conflict with interests that are identified within the scriptural sources (Qur'an and Sunnah) or the consensus of Islamic scholars (*'ijma*), and it cannot undermine a higher priority objective of Islamic law (Kamali 2003; Nyazee 2005). Furthermore, *maṣāliḥ* are further classified into three categories based on the textual linkage they have—*mu'tabarah*, *mursalah*, and *mulghāh*. *Al-Maṣlaḥah al-mu'tabarah* refers to those public interests that are expressly upheld by scripture since a law has been legislated within the texts to realize that interest. For example, the protection of dignity and honor is identified as the *ratio legis* for punishments prescribed in the scripture for slander and adultery. These interests are valid for the grounding of moral obligations because the "fact that the Lawgiver has upheld them is tantamount to His permission and approval of all measures, including legislation, that aim at their realization" (Kamali 2003). *Al-Maṣlaḥah al-mursalah* refer to interests that are not found within the Qur'an and Sunnah, and schools of Islamic law disagree over whether they are sufficient evidence for a moral obligation. The final type of *maṣlaḥah* is *maṣlaḥah mulghāh.* These interests run counter to scriptural source texts (Kamali 2003).

From the preceding discussion, one can assert that for the provision of healthcare to be an Islamic obligation, providing healthcare must secure a benefit that has been noted within the scripture, in other words a *maṣlaḥah mu'tabarah,* that results from providing healthcare must be identified. The obvious candidate *maṣlaḥah mu'tabarah* is life and the candidate *maqṣad* (singular form of *maqāṣid*) is the preservation of life, *ḥifẓ al-ḥayāt* or *ḥifẓ al-nafs*. Securing life (collective/societal) is a public benefit legitimated by the Qur'an as it is a *ratio legis* behind some of the punishments mentioned therein. While a healthcare system focused on clinical care does help to preserve life, research notes that only ten percent of one's health status and health outcomes are attributable to clinical care. Rather, the majority of one's health is determined by one's genetic predisposition for disease, the risks posed by or protective influences of the social environment he/she resides within, and one's dietary and exercise habits (Braveman and Gottlieb 2014; McGinnis et al. 2002). Hence, claiming that the provision of a hospital-centered medical care system is an Islamic moral obligation because it accords with the scripturally sanctioned public benefit of life (collective, societal not individual) must be made cautiously because such a system contributes to the protection of life only in a partial way.

Returning back to our consideration of whether the provision of healthcare is a moral obligation that emerges from the category of *ḥuqūq Allah*, the provision of healthcare must also be tied to essential (*ḍarūrī*) aspects of one of the five higher objectives (*maqāṣid*) of Islamic law. With respect to the higher objectives of Islamic law, al-Shatibi, the thirteenth-century legal scholar, who was one of the first jurists to lay out a comprehensive theory of the *maqāṣid,* notes that the essential aspects of the higher objective of the preservation of human life involve "establishing its (human life's) foundation through the legitimacy of procreation, ensuring its survival…by providing food and drink," and by "providing clothing and shelter" (Raysūnī and International Institute of Islamic Thought 2005). While al-Shatibi does not specifically mention the provision of medical care as aspects of *ḥifẓ al-ḥayāt,* it appears that the social determinants of health, e.g., food, shelter, etc., can be encompassed within his classification schema. As such at least certain societal conditions that promote health are public interests that are confirmed by Islamic law, and the absence of which can be deemed to impinge upon the rights of God. Muslim rulers are therefore obligated to provide these social goods.

It appears then that the provision of a comprehensive public health system—one that helps to preserve the social determinants of health—may be readily argued for on the basis of a traditional conceptualization of the essential aspects of *ḥifẓ al-ḥayāt.* And since one can furnish the argument that life, as a *maṣlaḥah mu'tabarah,* is enhanced to a greater degree by a public health system than a healthcare system primarily focused on clinical care, the provision of public health may be the greater Islamic obligation.

Finally, it bears mention that contemporary scholars have (re)interpreted *ḥifẓ al-ḥayāt* to also entail the "protection of certain parts of the body from harm or damage" and the "protection against infectious diseases" and go so far as to equate this objective with the Western bioethical principle of the right to life (Attia 2007; Auda 2008). Under this more expansive definition, a more direct case can be advanced for the nondiscriminatory provision of clinical healthcare to be an Islamic responsibility for the Muslim state.

Before moving to a discussion of communal obligations and healthcare, I would like to add a further comment regarding the structure of healthcare provision from an Islamic ethico-legal vantage point. As mentioned above, one of the distinguishing features of *ḥuqūq Allah* is that public interests and communal benefits are at stake. A healthcare system that does not deliver services to the general populace, i.e., is restricted to a certain group of individuals, would challenge the notion that a general public interest is at stake and that a communal benefit is accruing. In other words, if only a certain group of citizens are provided healthcare, the ethico-legal argument noted above may fall apart since only private interests are advanced. Hence the nondiscriminatory nature of healthcare provision is critical to make an Islamic argument for the state's provision of healthcare.

14.4 Considering the Provision of Healthcare to Be a Communal Islamic Obligation

In the discussion above, I considered whether the provision of healthcare is an Islamic obligation upon the state and whether it can be classified as part of *ḥuqūq Allah*. I now discuss the provision of healthcare as part of *ḥuqūq al-'ibād* and whether it can be deemed a *farḍ al-kifāyah* since these two Islamic ethico-legal constructs are close analogues to the concepts of social responsibility and communal obligation. Looking to the scriptural sources of Islam, there are a multitude of Qur'anic verses and Prophetic traditions that relate to social responsibilities and communal obligations in the context of health and healthcare.

Visiting the sick is classified as part of *ḥuqūq al-'ibād* and is also a communal obligation—a *farḍ al-kifāyah* (al-Munajjid). Speaking directly to this point, the Prophet Muhammad is reported to have said "the rights of one Muslim over another are six… If you meet him, greet him with salaam; if he invites you, accept the invitation; if he asks for advice, give him sincere advice; if he sneezes and praises Allah, say may Allah have mercy on you, if he falls sick, visit him; and if he dies, attend his funeral" (al-Nawawī). Another narration clarifies this obligation further where the Prophet reports of God's questioning humankind on the Day of Judgment saying "O son of Adam, I was sick and you did not visit me" and the human responds "O Lord how could I visit Thee whereas Thou art the Lord of the worlds?" to which God's response is, "Did you not know such and such servant of Mine was sick but you did not visit him and were you not aware of this that if you had visited him, you would have found Me by him?" (ibn al-Hajjaj Hadith 2569). These textual evidences support the idea visiting the ill is a sacred communal obligation upon the Muslim community.

While visiting the ill is a shared responsibility of Muslim community members, is providing cure also a communal obligation? This question has been addressed by Islamic legists in light of the scriptural sources as well. Both classical and modern Islamic jurists have classified the science of medicine as part of the knowledge required for the well-functioning of Muslim society and therefore studying medicine is a *farḍ al-kifāyah* (Ghaly 2010). One of the rationales behind this determination is that the restoration of health to the ill enhances their ability to lead beneficial and productive lives. The Qur'an states: "I (God) created the jinn and humankind only that they might worship Me (51:56)." (Pickthall 1938) Since the private and public worship—worship defined as both prayer and performing good works—is the *raison d'être* of creation, it stands to reason that a healthy individual is more capable of worship than one who is ill. A prophetic tradition conveys this notion and states "A strong believer is better and is more lovable to Allah than a weak believer, and there is good in everyone" (Ibn Mājah Vol. 1, Book 1, Hadith 79). Linking these two scriptural references together Islamic scholars explain that a strong (mentally, physically, and spiritually) human is more beloved to God because he is better able to discharge religious obligations and benefit oneself and the society at large. Consequently, health practitioners who can help the ill to regain health are deemed essential for the well-functioning of a Muslim society.

Yet there is a difference between the communal obligation to visit the ill and the *farḍ al-kifāyah* of providing healthcare. Notwithstanding chaplaincy visits, the communal obligation to visit the ill can theoretically be carried out by individuals without any special expertise. Healthcare, on the other hand, requires specialized knowledge and training, and meeting the health needs of society requires institutions. To be sure, the provision of healthcare requires a coordinated, systematized approach that links institutions such as medical schools that train clinicians, hospitals and rehabilitation facilities that can house the ill as well as provide specialized treatments, pharmacies that can dispense medications, research institutions that can develop novel therapeutics and systems of care, and licensing and certification bodies that can assess the quality of healthcare that is provided to one another. While individuals or other private entities could band together to provide this sort of healthcare system, in the modern era the provision of healthcare often requires the financial, bureaucratic, and human resources of the state. While a *farḍ al-kifāyah* can be discharged by a state apparatus, the discharge of a communal responsibility by state authorities is not the same as stating that it is an Islamic obligation for Muslim state authorities to provide healthcare to the citizenship. The former is discharging the responsibility on behalf of another party, while the latter suggests that it is the state itself that is liable. While it indeed may be an Islamic moral responsibility for the Islamic state to provide for the health needs of its citizens, additional concepts of Islamic law beyond *farḍ al-kifāyah* and *ḥuqūq al-'ibad* and other scriptural source texts may be needed to support that argument.

14.5 Enhancing Healthcare by Attending to Religious Needs

Educational curricula, therapeutic modalities, healthcare technologies, and healthcare personnel all move across national boundaries in the current era of rapid healthcare globalization. As a result, healthcare systems in vastly different sociocultural contexts have come to more closely resemble one another in structure and function, and the bioscientific models of clinical practice have become the dominant ways of attending to human illness and disease. Largely lost in this homogenizing movement has been due attention to the ways in which religion impacts understanding of health and disease and influences health behaviors, as well as developing systems of care that attend to the religious needs of patients. Assuredly, at least in Western nations, healthcare chaplaincy has been injected into the healthcare system in order to remedy the inattention to the spiritual needs of patients. However, chaplain staffing is often deemed inessential and is being cut in order to control costs, and, by and large, Muslim chaplains are rare and the religious needs of Muslim patients are especially underappreciated and frequently unaddressed (Abu-Ras and laird 2010). In the Muslim world, the situation is not much different as healthcare chaplaincy, at least in its nondirective, nondenominational form, is a practice that is somewhat alien to Muslim culture and thus the core knowledge and skill competencies of Islamic chaplaincy have yet to be formulated.

Against this backdrop there have been several initiatives in which Muslim stakeholders have sought novel solutions to attend to the religious needs of Muslim patients in the hospital. One US example is mosque-based patient visitation committees. In 2009, I conducted 13 focus groups with mosque-attending Arab, South Asian, and African American Muslims in Southeast Michigan to better understand the ways in which Islam impacted their health beliefs and healthcare-seeking behaviors (Padela et al. 2011). During those interviews I found that several mosques operated volunteer committees comprised of individuals who could frequent local hospitals and visit admitted Muslim patients in those facilities (Padela et al. 2012). During these visits community members would pray with patients, engage in social conversation, and offer a listening ear and at times even attended to the dietary needs of the ill Muslims. While visiting ill community members is not unique to Muslims, the practice is motivated out of a sense of a communal religious obligation, *farḍ al-kifāyah,* and as one of the rights Muslim community members had over one another, *ḥuqūq al-ibad.* Indeed focus group participants noted the religious teachings regarding the visitation of the ill as part of their Muslim identity and remarked that the dearth of Muslim chaplains and the unfamiliarity that most American healthcare practitioners had with Islamic mores and religious rites made the practice all the more important.

Moving from a Muslim minority context to a Muslim majority one, and from a mosque community-based practice to one that is institutionalized in hospital culture, is the advent of worship-friendly hospitals in Malaysia. Like many countries, the healthcare system in Malaysia is patterned after European/American designs. However in late 2009/early 2010, the International Islamic University of Malaysia (IIUM) received government approval to create a new teaching hospital. The leadership of IIUM sought to create a facility that is "able to support—holistically—the spiritual, mental and physical" aspects of patients such that they are "constantly... reminded of the Creator and always in the state of doing good deeds or *ibadah* (worship)" (Nawawi et al. 2012). To realize this objective, university representatives researched the designs of medieval Muslim hospitals, *bimaristans*, consulted Islamic legal experts so as to inform the hospital design based on Islamic ethico-legal injunctions regulations regarding patient care, religious rituals, and financial operations; and consulted with architects over a 2-year period. Construction of the hospital began in July 2012, and the hospital is expected to be fully functioning by 2016. According to the director of hospital management, the facility will be a model modern hospital for the Muslim world that "will conform to the shari'ah compliance and (the) *ibadah* friendly concept" (Zulkifly 2014) and will incorporate these principles in the standard operating procedures and daily activities of the hospital. Accordingly, the procedures and services provided by clinical staff will not contradict Islamic law, there will be adequate space and policies that enable staff and patients to perform the ritual prayers, patient and physician dress codes will allow for the following of the Islamic regulations, and staff will be trained so that they can advise patients on how to pray or fast when ill and/or whether they are exempted from praying and fasting due to illness. While this hospital has yet to open, a smaller, private hospital in Malaysia, Al Islam Specialist Hospital, has also implemented a worship-friendly ethos. Similar to the IIUM's objective, Al Islam seeks to be an environment that facilitates worship and educates patients and families about religious coping. Patients are assisted by trained personnel about the regulations and considerations of prayer and fasting while ill, and upon admission each patient is provided with a kit that contains material to facilitate worship such as a *qibla* finder (a compass that points toward Mecca which is the direction Muslims turn to pray), a parcel of clean dust for *tayammum* (clean dust is used to perform ritual ablution if one cannot use water), and head coverings. Furthermore, the nursing staff will assist patients in making ablution and reminding them of prayer request. By incorporating these services, the hospital focuses on the well-being of patients by assimilating worship during clinical treatment. The worship-friendly hospital movement also appears motivated out of Islamic sensibilities as it illustrates the keen interest Muslim groups have in actualizing the end goals of creation, *worship*, even while in hospital and represents a significant embellishment upon conventional hospital design and healthcare delivery. While the cultural competency movement suggests that clinical staff approach patients with humility and develop cross-cultural communication skills in order to minimize the conflict over cultural practices and values, the worship-friendly hospital goes one step further in training staff to actively facilitate worship (e.g., by helping patients perform ablution) and offer religious advice.

These two examples, the first of a patient visitation committee, and the second of the construction of healthcare delivery system designed to facilitate the religious lives of patients, while authentically Islamic from the perspective of the individuals involved, is not restricted to Muslim sensibilities. In other words the motivation to visit members from one's community who are hospitalized, or the desire of a patient to be visited by people from one's own background, is not something unique to Muslims; rather these are social needs and desires that are common to humanity. Similarly the desire of a particular community to build a hospital that is attuned to the social mores of that community while at the same time facilitating social habits of daily life that are important to that community is a notion that is shared across the religious and cultural divides of humankind. Consequently, detailed studies into the social lives of religious communities may offer insight into the shared impulses and needs of humankind and complement the philosophical inquiries into common morality and universal norms.

14.6 Final Remarks

Good health is an essential prerequisite for optimal human flourishing, and healthy populations are integral for the well-functioning of society. While people may differ in their conception of what constitutes health and how societies should be ordered, the aforementioned two axioms are beyond dispute. The UNESCO Declaration on Bioethics and Human Rights sets out a moral imperative for governments to improve the social determinants of health and to provide for the healthcare of its citizens through the ethical frameworks of social responsibility and universal human rights. While Muslims would agree with the necessity of health for optimal human flourishing and the critical importance of healthcare systems for societal functioning, the ethico-legal frameworks that undergird the Islamic moral obligations for healthcare provision are distinct from the ethical principles utilized by the UNESCO Declaration. This paper attempts to foster greater intercultural dialogue on the moral frameworks underpinning the "Islamic" right to health and healthcare.

This essay also presents two real-life examples of how Islamic ethico-legal concepts motivate Muslim behavior regarding care of the ill. These case narratives allow for moving beyond theory to practice and call attention to the potential benefits of descriptive studies motivating religious community practices. Such studies may bring to light novel approaches to health promotion and healthcare delivery that resonate with individuals outside of that particular religious community and offer strategies that can utilized more broadly to deliver holistic healthcare.

Acknowledgments Parts of this work have been previously published in as "Social Responsibility and the State's Duty to Provide Healthcare: An Islamic Ethico-Legal Perspective" by the same author in *Developing World Bioethics* 2017. The citation is provided in the reference list below.

I would like to thank Fr. Joseph Tham and Dr. Alberto Garcia Gomez and the UNESCO Chair for the kind invite to participate in the conference that motivated this paper. I also want to acknowledge Dr. Dan Sulmasy for his careful review and editorial comments on a previous version of this manuscript, Taha Abdul-Basser for his review of the Islamic ethico-legal reasoning and concepts described herein, and all the colleagues and interlocutors at the UNESCO meeting who spurred critical revisions of this essay.

References

Abdur-Rashid, K., S.W. Furber, and T. Abdul-Basser. 2013. Lifting the veil: A typological survey of the methodological features of Islamic ethical reasoning on biomedical issues. *Theoretical Medicine and Bioethics* 34 (2): 81–93.

Abu-Ras, Wahiba, and L. Laird. 2010. How Muslim and non-Muslim chaplains serve Muslim patients? Does the interfaith chaplaincy model have room for Muslims' experiences? *Journal of Religion and Health* 50: 46. https://doi.org/10.1007/s10943-010-9357-4.

al-Munajjid, Muhammad Saalih. *The rights of one Muslim over another include those that are obligatory and those that are mustahabb.* http://islamqa.info/en/178639. Accessed 9 Sept 2014.

Riyad as-Salihin. Sunnah.com. http://sunnah.com/riyadussaliheen

Attia, Gamal E. 2007. *Towards realization of the higher intents of Islamic law*. Herndon: The International Institute of Islamic Thought.
Auda, Jasser. 2008. Maqasid Al-Shariah: A beginner's guide. In *Occasional Papers Series*. London.
Braveman, P., and L. Gottlieb. 2014. The social determinants of health: It's time to consider the causes of the causes. *Public Health Reports* 129 (Suppl 2): 19–31.
Cairo Declaration on Human Rights in Islam. 1990. Cairo Egypt: Organization of Islamic Conference.
International Bioethics Committee. 2010. *On social responsibility and health*. Paris: United Nations Educational, Scientific and Cultural Organization.
Emon, A.M. 2006. Huquq Allah and Huquq Al-Ibad: A legal heuristic for a natural rights regime. *Islamic Law and Society* 13 (3): 325–391.
Fard al-Kifayah. 2003. In *The Oxford Dictionary of Islam*, ed. John L. Esposito: Oxford University Press.
Ghaly, Mohammed. 2010. *Islam and disability: Perspectives in theology and jurisprudence*, Routledge Islamic studies series. London/New York: Routledge.
Sunan Ibn Majah. Sunnah.com http://sunnah.com/ibnmajah
Kamali, Mohammad Hashim. 2003. *Principles of Islamic jurisprudence*. 3rd rev. and enl. Aufl. Cambridge: Islamic Texts Society.
McGinnis, J.M., P. Williams-Russo, and J.R. Knickman. 2002. The case for more active policy attention to health promotion. *Health Aff (Millwood)* 21 (2): 78–93.
Moosa, E. 2004. The Dilemma of Islamic Rights Schemes. *Worlds& Knowledges Otherwise*.
Nawawi, Norwina Mohd, Abdul Razak Sapian, Mohammed Fauiza Abdul Rani, and Wan Zainuddin Wan Ismail. 2012. *The shaping of an ibadah friendly hospital: learnings from the bimaristian of the evergreen past and the practicality of today's demands*. Paper presented at the 8th World Congress and Exhibition on Design and Health, Kuala Lumpur, Malaysia.
Nyazee, Imran A.K. 2005. *Theories of Islamic law*. Islamabad: Islamic Research Institute International Insitute of Islamic Thought.
Padela, A.I. 2017. Social responsibility and the state's duty to provide healthcare: An Islamic Ethico-legal perspective. *Developing World Bioethics* 17 (3): 205–214.
Padela, A., K. Gunter, and A. Killawi. 2011. *Meeting the healthcare needs of American Muslims: Challenges and strategies for healthcare settings*. Washington: Institute for Social Policy & Understanding.
Padela, A.I., A. Killawi, J. Forman, S. DeMonner, and M. Heisler. 2012. American Muslim perceptions of healing: Key agents in healing, and their roles. *Qualitative Health Research* 22 (6): 846–858. https://doi.org/10.1177/1049732312438969.
Pickthall, Mohammed Marmaduke. 1938. *The meaning of the glorious Qur'an*. Beltsville: Amana publications.
Raysūnī, Aḥmad, and International Institute of Islamic Thought. 2005. *Imam al-Shatibi's theory of the higher objectives and intents of Islamic law*. London/Washington: International Institute of Islamic Thought.
Sachedina, Abdulaziz Abdulhussein. 2009. *Islam and the challenge of human rights*. Oxford: Oxford University Press.
Sahih Muslim Book 7 Hadith 896.
Sahih Muslim Hadith 2162.
The Universal Declaration of Human Rights. 1948. Paris, France: United Nations.
United Nations Educational, Scientific and Cultural Organization. 2006. *Universal declaration on bioethics and human rights*. Paris: United Nations Educational, Scientific and Cultural Organization, Division of Ethics of Science and Technology, Social And Human Science Sector.
Universal Islamic Declaration of Human Rights. 1981. Islamic Council of Europe.
Yacoub, A. 2001. *The Fiqh of medicine*. London: Ta-Ha Publishers Ltd.
Zulkifly, Ahmad H. 2014. IIUM teaching hospital: The way forward? *The International Medical Journal of Malaysia* 13 (1): 1.

Chapter 15
Contextualizing the Islamic Perspective on Social Responsibility: A Response to Aasim Padela

Nouzha Guessous

Abstract The sacredness of human life is shared by all religions and cultures, despite existing differences in the content and value that individuals and societies attribute to the idea of being healthy or sick. Being considered as the text of God, the Qur'an is the main source of moral and ethical principles that should govern personal and social life of Muslims. Thus, its understanding and interpretation has a major and decisive impact on the rules of life as well as laws in most if not all Muslim states and societies. The approach based on a textual/wording interpretation of the commandments and prohibitions of the Qur'an and other derived historical texts may not take into account the context in which it was revealed or elaborated. Therefore, a pragmatic and contextualizing approach based on anthropology and sociology is more adapted to the evolving needs and aspirations of Muslims all over the world. Respect for human life includes the duty of taking care of ills, protecting health for all, and prohibiting "doing harm" to others. The Islamic religious duty of protecting human life implies that Muslim state's governments have the responsibility to care for the ill, to provide equal access to healthcare services and to the conditions that are necessary to be in a good health like adequate clothing, housing, food, and environment, as well as the protection of personal security, as stated in Art 14 of the UNESCO Universal Declaration on Bioethics and Human Rights.

15.1 General Introductory Comments

My understanding of the initiative of UNESCO Chair in Bioethics in Rome in organizing such workshops and conferences is to promote and improve the implementation of the Universal Declaration on Bioethics and Human Rights (UDBHR) by opening cross-cultural and interreligious debates on the principles of Bioethics proclaimed in the Declaration. Although the UDBHR is not a binding instrument, its interpretation and implementation may vary from a context to another, especially from a culture to another.

N. Guessous (✉)
Hassan II University, Casablanca, Morocco

J. Tham et al. (eds.), *Religious Perspectives on Social Responsibility in Health*, Advancing Global Bioethics 9, https://doi.org/10.1007/978-3-319-71849-1_15

These considerations all combined were clearly underlying the decision of the IBC to work on a Declaration of ethical principles common to all Humans as Humans which legitimate their global scope. After large discussions and hearings of representatives of most religions, philosophies, and cultures, the principles proclaimed in the UDBHR were unanimously adopted as Universal by all state members of UNESCO. This implies a set of rights and duties based on moral values shared by all humans and supported by all religions. Whether these rights are fulfilled or not is another issue depending on place, time, and circumstances. From this perspective, respect of human dignity in its dimension of rights and duties can and should take precedent over traditional/cultural – and even moral – prohibitions and obligations.

It is obvious that the Qur'an has a crucial importance in Islam as it is considered as the text of God. Therefore, it is the main source of moral and ethical principles that should govern personal and social life of Muslims. As a direct consequence, the understanding and interpretation of the Qur'an text has a major and decisive impact on the rules of life as well as laws in most if not all Muslim states and societies. Consequently, different schools of thoughts in Islamic Theology and Jurisprudence (Fiqh) emerged early in the seventh century AD and were followed by Islamic countries. The historical and current diversity of Muslim societies is one of its concrete consequences.

Padela's approach and conclusions are based on *one out of* other existing interpretations of the text of the Qur'an and other historical texts including some Islamic corpus of law. Other Muslim researchers and scholars endorse different approach and understanding of Islamic philosophy of life, based on the founding principles of Islam and on the evolution of Muslim's life and practices. I follow this direction of thought especially nowadays with the globalization and the worldwide intercultural connectedness.

The approach based on a textual/wording interpretation of the commandments and prohibitions of the Qur'an and other derived historical texts may not take into account the context in which it was revealed or elaborated. In other words, I believe that a pragmatic and contextualizing approach based on anthropology and sociology is more adapted to the evolving needs and aspirations of Muslims all over the world.

15.2 Islamic Versus Universal Instruments

In his very academic and documented paper, Padela calls on the Islamic corpus of law to conclude nevertheless that in fine, Islamic instruments such as the *Declaration of Human Rights in Islam* and the *Universal Islamic Declaration of Human Rights* share much in common with the end goals o*f Universal Declaration of Human Rights* (UNHR).

Even if in my opinion and appreciation, these Instruments are not so much known by Muslims, I agree with Padela that "they may have significance for the cultural acceptance of human rights by Islamic authorities." The real politics shows daily that each Muslim state is governed by its own bodies and has its national law and authorities. This makes the opinion of these "Supranational Islamic authorities" not binding for all Muslim states in fine; and the impact or even the reference to these Islamic instruments in National Constitutions and corpus of laws is neither systematic nor exclusive; it varies in space and time according to political regimes, governments, and other stakeholders and societies. It is important to highlight the fact that, without being secular, most if not all Muslim states agree on and endorse the principles of human dignity, justice, beneficence, nonmaleficence, and freedom in their National Constitution.

For example, the 2011 Moroccan Constitution[1] proclaims that "Islam is the religion of the state of Morocco" (Art. 3).[2] Compared to most of other Islamic State Constitutions, the Moroccan formulation does not specify that Islam is "the source of legislation," "the sole source of legislation," or "the foundation source of legislation." The monitoring of the Islamic identity of the laws is the responsibility of a *Council of the Ulama* (theologians) nominated by and under the supervision and authority of the King of Morocco as the "Commander of the faithful" (*Amir Al Mouminine*). In its preamble,[3] the 2011 Moroccan Constitution calls on the UN corpus of human rights. In article 19 Morocco has committed itself to "international conventions and pacts duly ratified, with respect for the provisions of the constitution, of the permanent characteristics and of the laws of the kingdom." Moreover, reference is made to the supremacy of international treaties over domestic legislation[4] in the preamble.

15.3 The Notion of Human Rights in Islamic Philosophy

I may agree partly with Sachedina (2009, 198, cited by Padela) assumption endorsed by Padela saying that "there is no unequivocal language of human rights in Islamic sources." However, I would add that this is also true in non-Islamic and Occidental worlds' acceptation and practice of the human rights in international politics and law, which is probably more important in terms of possible consequences.

For Padela, the concept of rights is different in Islamic philosophy from what it is in universal philosophy of Human Rights. According to his approach, he endorses

[1] http://www.constitutionnet.org/files/morocco_eng.pdf

[2] "Article 3 states that 'Islam is the religion of the state, which guarantees all the free exercise of beliefs.'"

[3] The preamble is stated as an integral part of the Constitution.

[4] "The Kingdom of Morocco...a) commits itself to comply with the international conventions duly ratified by it, within the framework of the provisions of the Constitution and of the laws of the Kingdom, within respect for its immutable national identity, and on the publication of these conventions; b) to recognize their primacy over the internal law of the country, and c) to harmonize in consequence the pertinent provisions of national legislation."

the concept of two primary kinds of rights: the "rights of God" and the "rights of humankind."

However, in his developments, one can understand that Islam do not recognize human rights as inherent to humans because and for they are humans. This is not coherent with the recurrence in the Qur'an of the notion of dignity of human beings, without discrimination of any kind, including for believers from other religions as well as non-believers. Furthermore, even from Padela's perspective, recognizing rights to humans does not neither overlap nor compete with the rights of God as duties for the believers.

In fine, inherent dignity provides the common moral and ethical foundational principle for all humans; and its subsequent protection duty is not specific or exclusive to any particular culture or religion. As such, it is universal.

15.4 Social Responsibility and Health in Islamic Philosophy

The sacredness of human life is common to all religions. It includes the duty of taking care of ills, protecting health for all, and prohibiting *doing harm* to others. As the well-being of everyone depends and impacts on the well-being of others and of the community, it is not only individual responsibility but also the responsibility of society, community, and state leaders.

From his reading of some Islamic sources, Padela suggests the concepts of *huquq al-ibad* and *fardh kifayah*, respectively, as partial analogues to the concepts of human rights and social responsibility/obligation, adding that these concepts have been used to support the Muslim states and community obligations to provide healthcare. Finally, even when using these concepts, he concludes that (1) the well-being of Muslim individuals and societies is a "right" and (2) the provision of healthcare is a communal Islamic obligation and a central purpose of governments and "Muslim rulers."

The sacredness of human life is shared by all religions and cultures, despite existing differences in the content and value that individuals and societies attribute to the idea of being healthy or sick. Over the centuries, the world's major religions have put health as individual's responsibility (UNESCO 2009). At the same time, all religions consider the preservation of health as an essential objective under the responsibility of the "community leaders." Since religions call for solidarity and empathy among humans, they have encouraged "the extension of the notion of responsibility for others beyond the confines of the family to communities, countries and across national boundaries" (Report of the International Bioethics Committee of UNESCO on Social Responsibility and Health 2010). Social responsibility should thereafter be understood as part of what has traditionally been called as human moral obligation.

The Qur'an puts the values of human dignity, justice, equality, and solidarity as essential in all areas of Muslim life and social organization. The preservation of human life is one of the five cardinal obligations that Muslims must protect: life (*An-Nafs/al-Hayat*), faith/religion (*Ad Din*), reason/mind (*Al A'ql*), filiation/lineage (*An Nasl*), and property and funds (*Al Mal*).[5]

The Islamic religious duty of protecting human life implies that Muslim state governments have the responsibility to care for the ill, to provide equal access to healthcare services and to the conditions that are necessary to be in a good health like adequate clothing, housing, food, and environment, as well as the protection of personal security,[6] as stated in Art 14 of the UDBHR.

I would like to give one interesting example of the power and the benefit of recalling on Muslim religious obligation to protect the life and health of Muslim citizens. In many Majority Muslim Societies (MMS), young girls under 18 years can and are forced by their parents to be married, and this has dramatic consequences for their future in general, including health complications and deaths and especially in cases of early pregnancy. By arguing the religious duty of protection of health and life for women and children and recalling on the state responsibility in this issue, some Islamic countries succeeded in changing their national legislation. In Morocco since 2004, marriage of minors under 18 years is prohibited, unless exceptional situation and under strict conditions controlled by the Family Court.

Other issues can benefit from the religious/cultural approach for the implementation of social responsibility as well as other principles of the UDBHR in a Muslim context. For example, one can argue that the protection of life and health is a responsibility and a duty of each believer. This means being accountable to God *and* to others for one's actions. Awareness campaigns can aim at reducing risky behaviors such as smoking, drugs and alcohol addiction, etc. by recalling the consequences on others and on society: secondhand smoking, delinquency social violence, traffic accidents, etc.

Other examples show how Islamic religious principles can positively be used for the implementation and promotion of the principles of the UDBHR in Muslim contexts. Religious acceptability of organ donation has been grounded, among other arguments, on the duty of saving lives, solidarity, and the protection of the community. These may have the priority over the will or even the interest of its individuals, including the management of health resources especially when they are limited. Appealing to social responsibility from its religious perspective may promote organ donation.

Other successful civil society initiatives approved and supported by official local and national religious authorities clearly showed that the involvement of religious leaders is very powerful. For example, awareness campaigns on AIDS (as well as on

[5] Each of these five objectives has three dimensions: the essential (*daruriyyat*), the necessary (*haji-yyat*), and the embellishment (*tahsiniyaat*).

[6] Aya 11 de Sourate Al Maida: ا لتق امنأكف ضرألا يف داسف وأ سفن ريغب اسفن لتق نم اعيمج سانلا ايحأ امنأكف اهايحأ نمو اعيمج سانل "http://library.islamweb.net/newlibrary/display_book.php?flag=1&bk_no=50&surano=5&ayano=32#docu

traffic accidents, violence, etc.) are yearly organized in Morocco with the participation of religious scholars in the media. Some imams of the mosques were asked to participate in these campaigns with their Friday's preaching. They promote prevention measures, such as chastity first and then use of condoms. They also call for empathy and non-stigmatization of AIDS patients.

Finally, I would like to add a few words on the Islamic grounding of the need to improve the social determinants of health as stated in Art. 14.

1. On the issue of reducing illiteracy and poverty, the Qur'an puts much importance on the benefits of education and access to knowledge for the believers. The first revealed Sourate of the Qur'an is "*Al Alaq*."[7] God makes the injunction to the prophet to read *('Iqra'e')*, which is understood as a duty for Muslims to be educated. Distributive justice and social solidarity toward the poor and the vulnerable is a Muslim obligation (e.g., the Zakat which is the ancestor of "State Taxes"). Interestingly, these two issues are linked in two versions of a Hadith from the prophet Mohamed: "Poverty and or ignorance may be blasphemous" (*Kada al faqr/al jahl/an yakouna koufran*).
2. On the issue of individual and societal responsibility toward the environment, it is interesting to recall that many verses of the Qur'an call on the duty to protect the future generations, other living organisms, and the environment as a whole. This could be an entry point for environmental ethics to be further developed by researchers.

15.5 Conclusion

The UNESCO Declaration of 2005 listed a set of Universal Principles in Bioethics and Human Rights after a long and interactive process of discussions and consultations involving both state and independent representatives of all cultures and religions. Nevertheless, implementation of these principles needs to be set in specific cultural as well as general context of each UNESCO Member State.

Cultural contexts may differ from a religious tradition to another. They may also differ within the same religious tradition, according to the specific history and to the current global context of each society and country. In some areas and under some circumstances, diversity among Muslim societies may be deeper than between Muslims, Christians, and Jews. There is an interesting diversity of opinions on the issues of bioethics and of human rights depending on the approach one takes when dealing with these issues from an *Islamic perspective*: the same texts can be interpreted with opposite meanings.

---

[7] Sourate N°96 "Al Alaq" (The Clot): (In the name of Allah, the Gracious, the Merciful) "Read! In the Name of your Lord, Who has created all that exists, Has created man from a clot (a piece of thick coagulated blood). Read! And your Lord is the Most Generous, Who has taught (the writing) by the pen, Has taught man that which he knew not."

In any case, the debate must continue among cultures and religions. Involving religious scholars and thinkers; as well as Social Sciences researchers and stakeholders; is necessary and beneficial for democratic decisions and for the promotion of the values of human dignity, equality, justice, autonomy, responsibility, and solidarity for the well-being and peace of all humanity.

References

Report of the International Bioethics Committee of UNESCO on Social Responsibility and Health. 2010. http://unesdoc.unesco.org/images/0018/001878/187899E.pdf

Sachedina, Abdulaziz Abdulhussein. 2009. *Islam and the challenge of human rights*. Oxford/New York: Oxford University Press.

The UNESCO Universal Declaration on Bioethics and Human Rights. 2009. Background, principles and application. In *Collective book*, ed. Henk Ten Have & Michèle Jean, 219–230. UNESCO. ISBN 978-92-3-104,088-7.

Chapter 16
The Universal Human Right to Access to Effective Health Care: A Catholic Christian Response to Aasim Padela

Stefan Kirchner

Abstract In Western, democratic societies, law and faith are usually kept distinct. Nevertheless did religious ideas also have a significant impact on the development of legal ideas in the Western world. This has been most notable in the development of human rights law based on Natural Law thought. Yet, human rights are universal and not dependent on specific cultural ideas, which is also the idea behind the universal accessibility of Natural Law. Today, human rights are defined in international documents, both binding treaties and non-binding but nevertheless notable, soft law documents. The right to access to effective health care is universal in nature and not dependent on citizenship, gender, age, faith, or other factors. While it was the Western world, which had been shaped by Christianity, which formed international law until long into the twentieth century, today's international law is based on the global community which no longer includes sovereign nations but encompasses everybody. This text looks at the Universal Declaration on Bioethics and Human Rights from a comparative cultural-religious perspective. The focus, though, is on the issue of universality and on the question how the universal approach found in Catholic Christianity has contributed to universality in modern human rights thinking, in particular when it comes to the right to access to effective health care. This then leads to the question of positive human rights obligations in the context of social responsibility.

Keywords Human rights · International law · Right to health · Social rights · Positive obligations

---

S. Kirchner (✉)
Arctic Centre, University of Lapland, P.O. Box 122, Rovaniemi 96101, Finland
e-mail: stefan.kirchner@ulapland.fi

J. Tham et al. (eds.), *Religious Perspectives on Social Responsibility in Health*, Advancing Global Bioethics 9, https://doi.org/10.1007/978-3-319-71849-1_16

16.1 Introduction

In particular in the context of Islam, it can be useful for outsiders to consider first the relationship between religiously motivated rules, religion, and ethical as well as legal norms. Similar to Christianity and the Jewish faith, Islam is a religion in which a written text plays a particularly important role. But while Jesus gave his followers the institution of a church rather than a book, the Qur'an is of a centrality to the Muslim faith which the Bible rarely holds for many Christians. Despite the very high level of the importance the Bible enjoys within Christianity, the importance attached to the Qur'an in Islam surpasses this in relative terms. For Christians, this might be explained by understanding that it is the Qur'an which is said to have been given from God to the faithful, while in the Christian faith, despite containing the word of God, the Bible is not the sole gift from God. For instance, Catholics believe that the institution of the Church and the sacraments are also fonts of divine inspiration.

When believers talk about the role faith plays for ethics and law, they will often have to do so in a highly secularized environment or at least one which has very little understanding for the idea that rules of behavior which are universal in character can be closely connected with religious concepts and values. Religiously motivated arguments regarding justice are often seen with suspicion.[1] This makes it imperative to learn about different approaches in different religions.[2]

16.2 The UNESCO Universal Declaration on Bioethics and Human Rights as International Law

The sources of international law, that is, public, rather than private, international law, are described in Article 38 (1) of the Statute of the International Court of Justice (Statute of the International Court of Justice 2014a):

> The Court, whose function is to decide in accordance with international law such disputes as are submitted to it, shall apply:
>
> (a) international conventions, whether general or particular, establishing rules expressly recognized by the contesting states;
> (b) international custom, as evidence of a general practice accepted as law;
> (c) the general principles of law recognized by civilized nations;
> (d) subject to the provisions of Article 59, [(Statute of the International Court of Justice 2014b)] judicial decisions and the teachings of the most highly qualified publicists of the various nations, as subsidiary means for the determination of rules of law (Statute of the International Court of Justice 2014c).

---

[1] See (Dabrock 2004, p. 20).

[2] For an overview over the Catholic position in bioethics, see (May 2008); for an excellent introduction to Bio*law* from a Catholic perspective, see (Palazzani 2002) and, (Palazzani 2009).

UNESCO's Universal Declaration on Bioethics and Human Rights (2014a) is not a binding treaty (Langlois 2008; Andorno 2007, 2013) but so-called soft law. Soft law, while not legally binding per se (Joyner 2005, p. 212), can inspire states and can reflect existing customary international law or lead to the emergence of new rules of customary international law.[3] The choice to phrase the Declaration in terms of has been criticized for being too European,[4] but - as we will see - the underlying premise of human dignity (which plays a key role in the Declaration[5]) is universal in nature and its protection is not tied to any particular culture or religion (Joas 2011, p. 21). The issue of universality can be baffling if one only thinks of an issue as being either ethics or law (Williams 2005, p. 215), as obliging either states or individuals (Williams 2005, p. 215), without taking into account the possibility of a more holistic set of obligations. With the increased awareness of individual responsibility and a greater role for soft law after 1945, international law has already evolved beyond such simple categories and today medical ethics, bioethics, is closely linked to the human rights (Faunce 2005, p. 177).

16.3 Universality of Human Rights: Claim or Reality?

But is such a document actually acceptable across cultural or religious fault lines? In other words, can religious values have a place in human rights norms which claim to be universal? The idea of universal human rights has its origins in Europe, specifically in a Europe which was still much more dominated by Christian, and specifically Catholic, ideas than this is the case today. As we will see in a moment, the particular Catholic emphasis on Natural Law[6] always has a certain degree of universality build in, which is also in line with the general understanding of Catholicism, a term which itself derives from the Greek term "καθολικός" (katholikos), which can be translated as "referring to the whole," in contrast to referring only to one ethnicity or region.[7] The Catholic church stood out at its beginning and in many ways still does so today,[8] by transcending ethnical boundaries.[9]

The idea that certain values are universal is certainly not new. At the same time was there always room for different local developments, which do not take away anything from the original message and content of the faith.

[3] (Joyner 2005, p. 212). Specifically for the possibility of the UNESCO Declaration becoming customary international law, see (Faunce 2005, p. 177).

[4] See (Andorno 2007, p. 152) and Andorno's rejection of this claim ibid.

[5] See (Häyry and Takala 2005, p. 231).

[6] See, e.g., (Compendium of the Social Doctrine of the Church. 2004).

[7] See also the development from Matthew 15:24 to Matthew 28:19. The Bible translations used for the creation of this text are The New American Bible and Biblia Sacra Vulgatae Editionis.

[8] See (Allen 2009, pp. 21 et seq).

[9] Galatians 3:28.

There may be differences in style in how a particular holiday is celebrated, e.g., in a predominantly Catholic culture like that in Mexico or in remote parts of Finland where the faithful may see a priest once a month or in a secret church in a country in which believers are persecuted by the authorities. Within the same rite, e.g., the Latin rite, what is commonly referred to as the Roman Catholic Church, the liturgy will be the same and even in case of a different rite the faith will be universal, independent of the local culture.

In a similar manner, universal human rights are also protected on the regional level through human rights instruments which place emphasis on different aspects. While, for example, the European Convention on Human Rights (2014) can only be understood if one takes into account that it was drafted by Western European states in 1949 and 1950, in the early years of the Cold War, one has to understand the particular importance of the Catholic faith in Latin America in order to understand the Inter-American human rights system, which in Article 4 (1) of the American Convention on Human Rights (2014) even includes an explicit reference to the right to life of the unborn child[10] which is not found in the European Convention. The African human rights system, on the other hand, places a much stronger emphasis on mutual obligations rather than exclusively on individual rights. Other regional developments, for example, in the central Asian states of the former USSR, the Arab world, or East Asia, have not yet reached the same level of legal development because there, unlike in Africa, Europe, or the Americas, individuals cannot make use of an international court in which they could actually sue states over human rights violations. Also, some regional developments fail to reach the minimum universal human rights standard outside the Universal Declaration of Human Rights. On the other hand has the idea of universal human rights been criticized – elsewhere – for failing to take into account other faiths.[11]

In the context of UNESCO's Universal Declaration, the use of the concept of universality has been criticized for not taking into account existing structural inequalities (Rawlinson and Donchin 2005, pp. 259 and 262). This charge overlooks that universal human rights do not require universally identical actions to realize these rights. By establishing "the highest attainable standard of health [as] one of the fundamental rights of every human being," (United Nations Educational, Scientific and Cultural Organization, Universal Declaration on Bioethics and Human Rights 2014b) the Universal Declaration on Bioethics and Human Rights clarifies that the right to health care has to be as effective as possible and that it is indeed a subjective right, not merely a vague goal or an objective obligation. The realization of this right can also require states to remove inequalities wherever they exist and limit access to effective health care.

[10] See (Paúl 2012).

[11] See the examples provided by (Faunce 2005, p. 177).

16.4 Social Responsibility As an Aspect of Christian Faith

For Padela, the obligation states undertake under Article 14 of the UNESCO Universal Declaration on Bioethics and Human Rights is based on the idea of social responsibility. Following with the clarification from the perspective of Islam, the obligations (and hence rights) follow not from social relationships but from divine authority. From a Christian perspective, rights and obligations can have a range of sources, including society in the widest sense of the term. The most fundamental human rights, however, and this is where our focus is, follow from the fact that we are human beings. As humans, we have rights and hence also obligations toward our fellow humans.

For Christians, in particular Catholics, this understanding applies universally, through Natural Law. In the early twentieth century, the idea of Natural Law was widely ignored by national lawmakers as well as in international law in favor of positivism and a nearly exclusive focus of man-made laws. After the horrors of the Shoa and World War II, Natural Law resurfaced in international law not only in academic discourse but also in international treaties and other documents (Römelt 2006, pp. 19 et seq), most notably in the emergency of international human rights law as a distinct set of legal rules and as an academic discipline. Yet, there is still widespread confusion as to what Natural Law actually is.

The catechism of the Catholic Church summarizes this teaching in numbers 1954–1960. In particular, it states that:

> The natural law, present in the heart of each man and established by reason, is universal in its precepts and its authority extends to all men. It expresses the dignity of the person and determines the basis for his fundamental rights and duties.
>
> Application of the natural law varies greatly; it can demand reflection that takes account of various conditions of life according to places, times, and circumstances. Nevertheless, in the diversity of cultures, the natural law remains as a rule that binds men among themselves and imposes on them, beyond the inevitable differences, common principles (Catechism of the Catholic Church 1997, nos. 1956 et seq).

From the perspective of international law, Natural Law "is also not identical to international law (although there are some overlaps) or the ius gentium of antiquity. While international law, of which the Convention is a part, arises out of the consent of the subject of the international legal order, Natural Law is based on the natural, inherent, connection which exists between all created beings, in this case, between all human beings. Because this connection is a consequence of the fact that humanity has been created – from the perspective of believers – by God. From a Christian perspective it is also God who is the ultimate reason for the existence of Natural Law. Yet, Natural Law also applies to non-believers. Therefore, it has to be independent of any specific faith, religion or world view and has to be accessible by everybody. The tools to understand the commands of the Natural Law are both reason and conscience."[12]

Being universal in nature, Natural Law is also universally accessible. "The Natural Law is 'written on the heart' of each and every human, regardless of religion, and which is therefore accessible, in other words, knowable, by all of use. When we follow our conscience, we do so because we have tapped into this knowledge of Natural Law, we know the law and act accordingly – because our conscience tells us to do so (Kirchner 2013a, p. 31)."

This understanding of Natural Law means that it can guide bioethical and biolegal rules. While in the context of the Universal Declaration on Bioethics and Human Rights the issue of consensus is relevant,[13] consensus is not a requirement for Natural Law. Rather than coming into existing after being agreed upon by everybody or by a majority, Natural Law is preexistent, coming from God, and can be discovered by everybody.

When compared to the approach in Islam, we can see that there are different understandings of the role of social responsibility. While from the perspective of Islam "[n]otions of social responsibility [...] are insufficient by themselves to ground an [...] obligation towards healthcare [but] require substantiation within the sacred source-texts of Islam," the Catholic approach appears to be the other way around: social responsibility is a consequence of the Natural Law which is accessible to everybody and hence valid without any form of codification, be it in religious or legal texts. The Natural law follows from our human nature and the fact that we are created by the one Creator God who Christians believe to be the triune God who has become man and as such has shared in our nature.

The distinction found in Islam between *huquq al ibad* and *huquq Allah* does not exist in Natural Law thinking in the same way, although similar results can be achieved. Rather, Natural Law includes both the fundamental rights of individuals and group rights.

A classical example for the latter can be found almost a century before the dawn of international law as it is understood since the Peace of Westphalia of 1648, when states became the key actors in international law. In the late 1530s, Francisco de Vitoria, who was professor for Moral Theology at the University of Salamanca in Spain, explained in a series of lectures the moral/legal status of the indigenous peoples of the Americas. At this time, what we refer today as international law still was part of moral theology as it were the monarchs who embodied the territories

[12] (Kirchner 2013a, p. 29), footnotes omitted. Note that the apparent contradiction to No. 1955 of the Catechism in this earlier text concerning the phrase "divine law" is based on a more narrow use of the terms "law" and "divine" in my 2013 text when compared to the Catechism and is by no means to be understood to mean a difference in substance, which, I would hope, is clarified later in my 2013 text when the God as the source of Natural Law is highlighted but which could be misunderstood if one were to look only at that particular passage. It can even be argued that God is the ultimate reason for human compliance with legal or moral norms, see (Hoerster 2009, p. 60) but see also ibid., p. 71.

[13] (Langlois 2008, p. 49); see also more generally (Andorno 2002, pp. 960 et seq).

(what we would today call "states"). Being bound by morals essentially limited the monarch in a way international law limits states today.

Natural Law, like modern understandings of human rights, both protects the individual and can have a collective dimension. By protecting public health, for example, by setting up hospitals and faculties of medicine at universities and regulating health insurance services or a public health system, the state both acts in fulfillment of obligations owed by everybody within it but also serves the rights of every individual. While a distinction between public and individual interest is possible, the distinction between public interest and fundamental human rights is much less clear because the protection of (individual and collective) fundamental rights is within the public interest. This does not mean that everything that is in the public interest requires a religious or moral obligation. In fact, in extreme cases of unjust societies, the perceived public interest or the interest of the majority can even run counter to Natural Law. To choose an extreme example: while it might be in the public interest to save money and while it would save money to the general public if the state would not provide aid to poor foreigners in need of help, denying aid to foreigners would run counter to Natural Law obligations which do not discriminate, e.g., on the basis of ethnicity. The same argument applies to the treatment of particular diseases. While it might be very expensive to treat rare or infectious diseases, for example, Ebola, Natural Law requires doing so – not only because of a risk to the general public but because of the human dignity of every human being. Therefore, there is also an obligation to provide all possible medical help in cases in which doing so would be very expensive but only benefit a single individual.

Natural Law does not require anything impossible, but it takes precedent over codified obligations. The fact that the state can act in fulfillment of obligations incumbent on everybody does not free the individual from all obligations. The state (or another form of organization, such as a charitable nongovernmental organization) is merely an instrument with which a group of individuals can fulfill their Natural Law obligations toward others. If the state fails to do so, for example, by not providing effective emergency aid to those in need, the original Natural Law obligation of the individual has not ceased. If there is no state structure, the obligation does not disappear. Like the *fardh kifaya* explained by Padela, Natural Law obligations can be fulfilled by the community – but they always also remain obligations of the individual and toward the individual human beings we are facing. The Good Samaritan[14] also was not a health-care professional, but he did what was within his possibilities to alleviate the suffering[15] of the other human being.

[14] Luke 10:29–37.

[15] On suffering, see (John Paul II 2014).

16.5 Concluding Remarks

The nature of obligation can differ between Christianity and Islam, with the latter seemingly placing more emphasis on the text of the Qur'an and the former placing more emphasis on the created nature of all humans and using the Scripture not so much as the source of obligation but as a tool for understanding how we should treat others. In a sense, the Christian approach provides some apparent parallels to the sources of international law mentioned at the beginning in that writings and teachings serve to increase our understanding of our obligations. This comes as no surprise if we keep in mind the different origins of both texts: for Muslims, the text of the Qur'an comes directly from God as dictated to Muhammed by the Archangel Gabriel (Jibril), while for Christians, the Bible is God's word but written by humans who were essentially telling the story of God's interaction with humankind. The Islamic approach therefore appears to be rather textual,[16] while the Natural Law approach looks first at the other human as a human being. That said, it has to be noted that this view is not shared by all strands of Christianity. In particular in some groups of the Protestant denominations, the text of the Holy Bible may take precedence over all other considerations. They often reject the idea of any group having the authority of interpretation. In such a context, the Natural Law ideas outlined here might not be familiar for all Christians.

The existing differences between Natural Law thinking and the Qur'anic approach do not prevent a practical cooperation in the same direction. Different understandings of the origins of social responsibility do not make cooperation in the health-care sector impossible. On the contrary, the importance given to works of charity (long a contentious issue between Catholics and Protestants[17]) in both Catholicism and Islam should actually facilitate cooperation in many cases. In particular in Europe, this can play an important role in the future. Many European states are home to large Muslim immigrant communities. Many Muslim immigrants and their children and grandchildren will grow old in predominantly secular European nations. Padela has emphasized the importance of taking into account the religious needs of patients. Religiously motivated health-care providers on the other hand might be able to be more sensitive to the religious needs of patients, even if they are not adherents of the locally dominant faith. In so far, a return to Christian values in the health-care sector could explicitly benefit also non-Christian patients.

Although biolaw and bioethics can have many sources, human dignity is at the heart of biolaw and bioethics.[18] A key objective of international human rights law is the protection of human dignity of all humans without discrimination (Guerra

[16] On problems associated with such an approach from a legal rather than a theological perspective, see (Kirchner 2011, pp. 66 et seq).

[17] On the Catholic perspective, see (Hartin 1999, pp. 57 et seq).

[18] See in more detail (Andorno 2009), reprinted as (Andorno 2013).

Martins 2013, pp. 31 et seq). The universality inherent in the Natural Law approach reinforces this notion. Increasing the awareness of Natural Law obligations concerning the social responsibility of the individual and of society at large will enhance the quality and accessibility of health care in general.

References

Allen, J.L. 2009. *Das neue Gesicht der Kirche – Die Zukunft des Katholizismus*. 1st ed. Gütersloh: Gütersloher Verlagshaus.

American Convention on Human Rights [Pact of San Jose]. 2014. http://www.oas.org/dil/treaties_B-32American_Convention_on_Human_Rights.htm. Last accessed 1 November 2014.

Andorno, R. 2002. Biomedicine and international human rights law: In search of a global consensus. *Bulletin of the World Health Organization* 80: 959–963.

———. 2007. Global bioethics at UNESCO: In defense of the universal declaration on bioethics and human rights. *Journal of Medical Ethics* 33: 150–154.

———. 2009. Human dignity and human rights as a common ground for a global bioethics. *Journal of Medicine and Philosophy* 34: 223–240.

———. 2013a. Global bioethics at UNESCO: The universal declaration on bioethics and human rights. In *Principles of international biolaw – Seeking common ground at the intersection of bioethics and human rights*, ed. Roberto Andorno, 1st ed. Brussels: Bruylant.

———. 2013b. Human dignity and human rights as a framework for international biolaw. In *Principles of international biolaw – Seeking common ground at the intersection of bioethics and human rights*, ed. Roberto Andorno, 1st ed. Brussels: Bruylant.

Catechism of the Catholic Church. 1997. *Completed and updated, with modification from the Editio Typica*, 2nd ed. New York/London/Toronto/Sydney/Auckland: Doubleday.

Compendium of the Social Doctrine of the Church. 2004. *In the context of international law Pontifical Council for justice and peace*. http://www.vatican.va/roman_curia/pontifical_councils/justpeace/documents/rc_pc_justpeace_doc_20060526_compendio-dott-soc_en.html, No.437.

Convention for the Protection of Human Rights and Fundamental Freedoms [European Convention on Human Rights]. 2014. European Treaty Series No. 5, http://www.echr.coe.int/Documents/Convention_ENG.pdf. Last accessed 1 November 2014.

Dabrock, P. 2004. Fundamentaltheologische Bioethik angesichts der Herausforderungen moderner Gesellschaft. In *Menschenwürde und Lebensschutz-Herausforderungen theologischer Bioethik*, ed. Peter Dabrock, Lars Klinnert, and Stefanie Schardien, 1st ed. Gütersloh: Gütersloher Verlagshaus.

Faunce, T.A. 2005. Will international human rights subsume medical ethics? Intersections in the UNESCO universal bioethics declaration. *Journal of Medical Ethics* 31: 173–178.

Guerra Martins, A.M. 2013. *Direito Internacional dos Direitos Humanos*. 1st ed. Coimbra: Almedina.

Hartin, P.J. 1999. *A spirituality of perfection – Faith in action in the letter of James*. 1st ed. Collegeville: Liturgical Press.

Häyry, M., and T. Takala. 2005. Human dignity, bioethics and human rights. *Developing World Bioethics* 5: 225–233.

Hoerster, N. 2009. Ist Gott unverzichtbar für die Moral? In *Wozu Gott? Religion zwischen Fundamentalismus und Fortschritt*, ed. P. Peter Kemper, A. Mentzer, and U. Sonnenschein, 1st ed. Frankfurt am Main/Leipzip: Verlag der Weltreligionen.

John Paul II. 2014. *Salvifici doloris*. http://www.vatican.va/holy_father/john_paul_ii/apost_letters/documents/hf_jp-ii_apl_11021984_salvifici-doloris_en.html. Last accessed 1 Nov 2014.

Joas, H. 2011. *Die Sakralität der Person – Eine neue Genealogie der Menschenrechte*. 1st ed. Berlin: Suhrkamp Verlag.

Joyner, C.C. 2005. *International law in the 21st century-rules for global governance*. 1st ed. Lanham: Rowman & Littlefield Publishers.

Kirchner, S. 2004. Relative normativity and the constitutional dimension of international law: A place for values in the international legal system? *German Law Journal* 5: 47–64.

———. 2011a. Faith, ethics and religious norms in a globalized environment: Freedom of religion as a challenge to the regulation of Islamic finance in Europe. *Baltic Journal of Law and Politics* 1 (4): 52–82.

———. 2011b. *Multidimensional origins of biolaw and bioethics and their impact on conceptions of human rights and human dignity-a European perspective*. Munich: GRIN Verlag GmbH 2015.

———. 2011c. Personhood and the right to life under the European convention on human rights: Current and future challenges of modern (bio-)technology. *University of Warmia-Mazury Law Review* 3: 44–58.

———. 2012a. Turning religious values into law through the language of human rights: Legal ethics and the right to life under the European Convention on Human Rights. *Baltic Journal of Law and Politics* 5 (1): 70–98.

———. 2012b. The personal scope of the right to life under article 2(1) of the European convention on human rights after the judgment in A, B and C v. Ireland. *German Law Journal* 13: 783–792.

———. 2013a. Natural law as Biolaw. *Jurisprudencija* 20: 23–39.

———. 2013b. Biomedicine, biotechnology and the European convention on human rights: Life before birth. *Law and Forensic Science* 6: 8–48.

———. 2015a. *Transcultural biolaw in a hyper-commercialized society*, forthcoming (2015).

———. 2015b. *Unbalanced on a slippery slope? A shift in the right to life jurisprudence of the European Court of Human Rights with regard to Poland*, forthcoming (2015).

Langlois, A. 2008. The UNESCO universal declaration on bioethics and human rights: Perspectives from Kenya and South Africa. *Health Care Analysis* 16: 39–51.

May, W.E. 2008. *Catholic bioethics and the gift of human life*. 2nd ed. Huntington: Our Sunday Visitor Publishing Division.

Palazzani, L. 2002. *Introduzione alla Biogiuridica*. 1st ed. Torino: G. Giappichelli Editore.

———. 2009. *Introduction to the philosophy of biolaw*. 1st ed. Rome: Edizioni Studium.

Paúl, A. 2012. Controversial conceptions: The unborn and the American Convention on Human Rights. *Loyola University Chicago International Law Review* 9: 209–247.

Pichon, J. 2006. Does the unborn child have a right to life? The insufficient answer of the European court of human rights in the judgment Vo v. France. *German Law Journal* 7: 433–444.

Rawlinson, M.C., and A. Donchin. 2005. The quest for universality: Reflections on the universal draft declaration on bioethics and human rights. *Developing World Bioethics* 5: 258–266.

Römelt, J. 2006. *Menschenwürde und Freiheit – Rechtsethik und Theologie des Rechts jenseits von Naturrecht und Positivismus*. 1st ed. Freiburg/Basel/Vienna: Herder.

Statute of the International Court of Justice. 2014a. http://www.icj-cij.org/documents/?p1=4&p2=2#CHAPTER_I. Last accessed 1 November 2014.

Statute of the International Court of Justice. 2014b. Article 59 of the Statute of the International Court of Justice reads as follows: "The decision of the Court has no binding force except between the parties and in respect of that particular case. http://www.icj-cij.org/documents/?p1=4&p2=2#CHAPTER_I. Last accessed 1 November 2014.

Statute of the International Court of Justice. 2014c. Article 38 (1) Statute of the International Court of Justice. http://www.icj-cij.org/documents/?p1=4&p2=2#CHAPTER_I. Last accessed 1 November 2014.

United Nations Educational, Scientific and Cultural Organization, Universal Declaration on Bioethics and Human Rights. 2014a. http://portal.unesco.org/en/ev.php-URL_ID=31058&URL_DO=DO_TOPIC&URL_SECTION=201.html. Last accessed 1 November 2014.

United Nations Educational, Scientific and Cultural Organization, Universal Declaration on Bioethics and Human Rights. 2014b. Article 14 UNESCO Universal Declaration on Bioethics and Human Rights. http://portal.unesco.org/en/ev.php-URL_ID=31058&URL_DO=DO_TOPIC&URL_SECTION=201.html. Last accessed 1 November 2014.

Williams, J.R. 2005. UNESCO's proposed declaration on bioethics and human rights – A bland compromise. *Developing World Bioethics* 5: 210–215.

Part VI
Judaism

Chapter 17
Jewish Religious Perspectives in the Israeli Healthcare System

Jonathan Halevy and Adina Halevy

Abstract Although the laws of the State of Israel are not based on the Jewish religious law (*Halakhah*) but on secular law enacted by the Israeli Parliament (*Knesset*), Jewish religious elements and values are deeply embedded in the decisions of the legislative bodies that concern the Israeli healthcare system. To this effect we review in this paper the contents and ways of implementation of five laws accepted by Israeli legislative based on Jewish Law principles. The laws are the National Health Insurance Law, 1994; the Patient's Rights Law, 1996; the Dying Patient Law, 2005; the Brain/Respiratory Death Law, 2008; and the Good Samaritan Law, 1998. In conclusion, Israel's legal code is secular but incorporates principles and perspective of Jewish law, especially in the areas of danger to life and the provision of healthcare to all. These principles and perspectives are in full keeping with Article 14 of the UNESCO Declaration on Bioethics and Human Rights. The five examples discussed bear testament to the manner in which Israel's legislative body, the Knesset, has encompassed these values within the laws of the modern State of Israel.

Keywords Jewish law (Halakhah) · Universal healthcare coverage · End of life

The Jewish religion, since Biblical times and through the generations, has developed a body of law known in Hebrew as Halakhah. Still evolving by means of continual rabbinical debate, the Halakhah provides the orthodox Jew with rulings on all aspects of life, from birth to the grave. Nevertheless, the legislation of the modern State of Israel is a separate entity based on secular foundations.

Let us examine briefly the character of Israeli law, which was based initially on a blend of the legal systems of the Ottoman and the British Empire, both previous rulers of the region. Except in one particular area – family life, marriage, and divorce being regulated for the Jewish population by Halakhah – Israeli laws, as they have evolved from rulings of the Supreme Court since 1948, are secular. (The majority of

J. Halevy (✉) · A. Halevy
Shaare Zedek Medical Center, Jerusalem, Israel
e-mail: halevy@szmc.org.il

J. Tham et al. (eds.), *Religious Perspectives on Social Responsibility in Health*, Advancing Global Bioethics 9, https://doi.org/10.1007/978-3-319-71849-1_17

Israeli citizens define themselves as secular.) However, as I propose to illustrate, Jewish religious elements and values are deeply embedded in the decisions of the legislative bodies that concern the Israeli healthcare system.

I would like to examine in detail five of the most important laws pertaining to the Israeli healthcare system.

17.1 The National Health Insurance Law, 1994

This sets forth the state's responsibility to provide health services for all citizens of the country. The law opens with the statement: "Medical insurance, under this law, shall be based on principles of justice, equality and mutual assistance." In full keeping with Article 14 of the UNESCO Declaration on Bioethics and Human Rights, through the four health maintenance organizations (HMO), all citizens of Israel are entitled to a very wide basket of services which includes:

- Medical diagnosis and treatment both at clinics and at the home of the patient.
- Preventive medicine and health education (i.e., early diagnosis of embryo abnormalities, vaccinations, and counseling for pregnant women, mothers, and the elderly).
- Hospitalization (general, maternity, psychiatric, and chronic).
- Surgery and transplant. If medical treatment is not available in Israel, treatment abroad will be covered.
- Preventive dental care for children.
- First aid and transportation to a clinic or hospital.
- Medical services at the workplace.
- Medical treatment for drug abuse and alcoholism.
- Medical equipment and devices.
- Obstetrics and fertility treatment.
- Treatment of injuries caused by violence.
- Medication, in accordance with a list issued by the Ministry of Health.
- Treatment of chronic diseases.
- Paramedical services (i.e., physical therapy, occupational therapy, etc.).

An additional note on two of the above:

Fertility treatment. Israel is perhaps the only country in the world which provides treatment from public funding until two children are born (not two cycles of treatment) – an expression of the great importance Jewish culture gives to the family unit.

Medication and devices. Modern medicine is characterized by the very rapid progress continually occurring in new technologies – both medications and devices for diagnosis and treatment. In order to reflect this, the Office of Technology Assessment of the Ministry of Health annually issues a call for senior doctors, medical school deans, and other professionals to submit suggestions of recent innovations which should be publicly funded. A public committee including senior physicians, health economists, bioethicists, and public representatives meets to prioritize these items, depending upon factors such as efficacy in saving and prolonging life, quality of life attainment, and cost. Thus, the basket available to all citizens is revised and updated, as required in the UNESCO Article 14, item 2.

However, since in this era of costly modern medicine no country in the world can supply its citizens through public funding everything that contemporary medicine can offer, the question arises: what are the ethical principles that should underlie the process of prioritization? The aim of the committee described above is to determine the best use of the limited resources that Israel, in common with all countries, can provide for its health services.

The use of scarce resources is a topic that arose in Halakhic discussions many centuries ago. The basic question addressed is whether society is an entity with its own interests or whether it is merely the sum of all the individuals it encompasses. If the latter is true, every individual has the right to expect that society will meet all his particular needs in full. If, however, it exists above and beyond all the individuals it comprises, then society has its own needs, ethical and legal standards which must take precedence over those of any single individual (Steinberg 2003).

According to Halakhah, society is, indeed, judged to be a separate entity, and discussions within the context of redemption of captives have ruled that society cannot override the present and future requirements of the entire community by investing a disproportionate amount of money to redeem a single prisoner. Society must take into account its own needs even if it may thereby harm or interfere with the needs of an individual (Bavli Talmud, Tractate Gittin 45a).

In keeping with this, the Israeli committee which determines the content of the medical basket must give priority to drugs and treatments that will answer the basic needs of all its citizens; although with limited funding available, this results in the inevitable exclusion of effective but expensive treatments required by some patients. (It is, of course, perfectly acceptable that patients who are able to finance such treatments from their own pocket may do so).

It should be noted that the public basket is vast and all medications that are overtly life-saving or prolong life of quality are included. Usually excluded for budgetary reasons are medications that prolong life to a very limited extent (mainly chemotherapy that prolongs life by a matter of weeks at the optimum) or new medications and technologies whose efficacy is still under debate in current literature and which may not eventually hold a permanent place in treatment.

The National Health Insurance Law provides Israeli citizens with universal health coverage. A system of progressive medical insurance premiums requires that low-income earners pay 3.1% of their income toward this, while the majority pays more, on the basis of 4.8% of income. It should be stressed that those receiving welfare payments are entitled to the services basket outlined above free of charge. The high quality of medical care available to all is reflected in statistics that place Israel high in the table of developed countries with a life expectancy in 2012 of

83.6 years for women and 79.9 years for men and an infant mortality rate of 3.4 per 1000 live births (Central Bureay of Statistics 2012).

Such healthcare coverage for all citizens, which has not yet been achieved by a large number of countries, including the USA, is primarily an expression of the priority Israel's legislation gives to social justice, a value Judaism holds dear.

17.2 The Israeli Patients' Rights Law, 1996

Passed in 1996, the Israeli Patients' Rights Law defines the right of every individual, irrespective of gender, ethnic origin, age, or religion, to receive quality medical care with dignity. In keeping with the move toward patient autonomy, emphasis was laid on the necessity of detailed informed consent from patients for any diagnostic or therapeutic procedure they undergo.

However, a religious consideration strongly influenced an additional aspect of this law. The original intention was to include an article which would also establish a legal right "to die with dignity." This may have involved withdrawal from life support systems and even assisted suicide, both of which are in negation of the Jewish tenet – held indeed in common with other faiths – which places the sanctity of life above all considerations. The obligation to save life is one of the cardinal principles of Judaism (Bavli Talmud, Tractate Sanhedrin 37a). Taking into account the beliefs of the considerable religious sector of the population, and thus expressing both democratic and Jewish values, the legislative body of the Knesset (Israel's Parliament) could not reach consensus on this point. The law was passed without dealing with this specific matter.

17.3 The Dying Patient Law, 2005

An issue of such major importance could not be ignored. Seven years later, after much further deliberation in the Knesset and in public media, a committee of 59 physicians, nurses, social workers, psychologists, rabbis, philosophers, and lawyers headed by one of Israel's leading bioethicists was appointed by the Minister of Health. Following 2 years of intense discussion, the Dying Patient Law 2005 was enacted to direct the manner of end of life care in terminal illness. I see this law as an epitomization of consideration for all sectors of the Israeli population: for the religious, it recognizes the sanctity of life and the belief that the soul is God's and at His will and, for the secular, the autonomy that each person has over his own body. It also expresses the right of the patient not to suffer. As the opening of the law states:

Goal and Fundamental Principle

Goal: a. This Law regulates the medical treatment of the terminally ill patient based on an appropriate balance between the value of the sanctity of life, the value of the individual's autonomous will and the importance of quality of life. b. This Law is based upon the values of the State of Israel as a Jewish and democratic state and on fundamental principles in the realm of morality, ethics and religion.

Fundamental Principle: In prescribing the medical treatment of the terminally ill patient, his medical condition, his will and the degree of his suffering are the exclusive considerations.

This law forms a comprehensive legal guide to all physicians who attend those dying of an incurable disease. First, it defines the "dying patient" as one whose condition must be assessed by two senior physicians, and it must be determined that he has less than 6 months to live because of an incurable disease which greatly erodes quality of life. The patient's autonomy is held in high regard: if an advanced medical directive (living will) has been made ruling out life support, dialysis, or any treatment which purely prolongs life without curing, it must be implemented. To ensure the fulfillment of living wills, a registry for such documents has been established by the Ministry of Health.

However Jewish ethics are also prominent –most Rabbinic authorities (Responsa Igrot Moshe Yoreh Deah, Part3 #132, Responsa Tzitz Eliezer Part 17#72:13.) forbid active euthanasia in any form, thus, for example, it is not permitted to disconnect life support systems to which the patient has been attached. Life is sacred, not within man's jurisdiction. Optimal palliative care must be provided to minimize suffering.

This law is another manifestation of how Israeli legislators try to balance the heterogeneous views held by the various main groups comprising Israeli society – secular Jews, orthodox Jews, Muslims, Christians, and Druze. The law was passed in the Israeli Parliament by an overwhelming majority from all political parties.

17.4 The Brain/Respiratory Death Law, 2008

Cadaveric organ donation is of maximal benefit to the recipient where the circumstance of death of the donor is brain death, usually a result of head trauma or major stroke. Of the approximately 40,000 deaths that occur annually in Israel, only a very small proportion, around 200, are brain deaths, the remainder being cardiac deaths. The Harvard criteria for pronouncing brain death were formulated in 1968 (Report of the Ad Hoc Committee of the Harvard Medical School 1968) – describing a situation where extensive and severe brain damage is well-documented at both clinical level (characterized by deep coma with no response to external stimuli, no spontaneous breathing, no pupillary and corneal reflexes, and support of heart and circulation by artificial means) and at laboratory level (flat EEG indicating no electrical activity of the brain). Modern technologies now enable us to assess brain death even more accurately, and there are also more recent requirements of no electrical activity under stimulation or no blood supply to the brain on ultrasound doppler (transcranial Doppler – TCD).

Most Jewish authorities today have joined with modern science in recognizing brain death as defined above, but there are rabbis who do not accept the concept and oppose organ donations from brain dead cadavers.

The Israeli Brain/Respiratory Death Law stipulates that in Israeli hospitals it is the prerogative of the patient (by completing an organ donor card during his lifetime) or his family to choose or reject the option of donating organs. However, should they choose against this, frequently because they do not recognize brain death as death, the medical staff is committed to continue supportive measures until cardiac death occurs.

The law further stipulates that only certain doctors who have undergone special training on the matter of brain death may determine whether this is such a case, and confirmatory laboratory tests stimulating the brain or demonstrating no blood supply (not merely a flat EEG, deep coma, and basic reflexes) are mandatory.

Thus we see that this law reflects the convergence of the individual's autonomy with the religious views of that segment of the population, i.e., the ultra-Orthodox (chiefly Jews but also of other religions or secular), who do not recognize brain death. The ethical principles deeply rooted in this section of the population are democratically preserved.

17.5 The Good Samaritan Law, 1998

In common with many countries, Israeli law requires that all those in immediate severe danger should receive assistance from those in the close vicinity. Following Christian tradition, this law is frequently known as "The Good Samaritan" Law (Good Samaritan Law). In most countries, this law provides legal protection to people who give reasonable assistance to those who are injured, ill, in peril, or otherwise incapacitated. In some cases, Good Samaritan laws encourage people to offer assistance, and this legal protection is intended to reduce a bystander's hesitation to act for fear of being sued or prosecuted for causing unintentional injury or wrongful death. In Israel, the law, passed in 1998, takes its name from the biblical source: "thou shall not stand idly by the blood of thy neighbor" (Leviticus 19:16). The Israeli law concurs with that of other countries in specifying that the person giving help cannot be sued and is entitled to reimbursement of expenses that may be incurred. However, there is an additional requirement here – the "savior" *must* assist any person whom he sees to be in severe immediate danger, when he is able to do so without endangering himself or another person.

In conclusion, Israel's legal code is secular but incorporates principles and perspective of Jewish law, especially in the areas of danger to life and the provision of healthcare to all. These principles and perspectives are in full keeping with Article 14 of the UNESCO Declaration on Bioethics and Human Rights. The five examples discussed above bear testament to the manner in which Israel's legislative body, the Knesset, has encompassed these values within the laws of the modern State of Israel.

References

A definition of irreversible coma: Report of the Ad Hoc Committee of the Harvard Medical School to examine the definition of brain death. JAMA 205(6): 337–340 (August 5,1968). https://doi.org/10.1001/jama.1968.03140320031009.

Bavli Talmud, Tractate Gittin 45a.

Bavli Talmud, Tractate Sanhedrin 37a.

Central Bureay of Statistics, 2012. Israel Statistics Bulletin, 2012, in www.cbs.gov.il.

Good Samaritan Law, in Wikipedia: https://en.wikipedia.org/wiki/Good_Samaritan_law.

Leviticus 19:16.

Responsa Igrot Moshe Yoreh Deah, Part3 #132.

Responsa Tzitz Eliezer Part 17#72:13.

Steinberg, A. 2003. *Encyclopedia of Jewish medical ethics*. Vol. 1, 45. New York: Feldheim.

Chapter 18
Reflections on the Universality of Healthcare in Israel: A Jewish Response to Jonathan and Adina Halevy

Samuel Weingerz Mehl

Abstract Even though Israel is a secular state and most of its citizens identify as secular, the social, cultural, and spiritual legacy of Judaism is very prominent. The *Bible* contains the essence of Jewish ethics and is the base that nurtures the Jewish spirit. The result of the union of the *Mishnah* and *Guemara* is the *Talmud*, which, along with the *Torah*, forms the basis for the development of the *Halakhah*. Judaism has a vibrant tradition in medical ethics. Some essential aspects of human life that are regulated by Jewish religion are life and death, sexuality, pain, and suffering due to illness. While Israel's legal code is secular, it incorporates principles and perspective of Jewish law, especially in the areas of danger to life and the provision of healthcare to all. I endeavored in this response to show that religion, especially the Jewish faith, can be an essential element of cultural pluralism in contemporary Israel. There is no opposition between religion and healthcare provision. In contrast, religion serves as a guide, orientation, and inspiration for conduct in modern society and protection of human rights. The United Nation's goals of working together to establish universal principles in ethics and human rights that guarantee respect for human life are laudable. The concepts of equality, justice, and respect for human rights possess incalculable value and many implications.

Keywords Judaism · Medical ethics · Universal principles · Human rights · UNESCO

18.1 Introduction

As a Jewish bioethicist living in Mexico, it is my great pleasure to respond to the Halevys' document from an international focus. I will begin discussing the UNESCO/UN's definitions of health and bioethics and then discuss how Judaism

S. Weingerz Mehl (✉)
Dr Manuel Gea Gonzalez General Hospital, Mexico City, Mexico
e-mail: weingerz@prodigy.net.mx

J. Tham et al. (eds.), *Religious Perspectives on Social Responsibility in Health*, Advancing Global Bioethics 9, https://doi.org/10.1007/978-3-319-71849-1_18

and adherence to international bioethical norms have led Israel to be a model of near-universal health coverage.

I would like to commence by mentioning excerpts of the Universal Declaration on Bioethics and Human Rights that "health does not depend solely on scientific and technological research developments but also on psychosocial and cultural factors [and] that a person's identity includes biological, psychological, social, cultural and spiritual dimensions." In the specific case of Israel, even though it is a secular state and most of its citizens identify themselves as secular, the social, cultural, and spiritual legacy of Judaism is very prominent.

It is essential that universal principles are established as a basis to respond to on the ever-increasing dilemmas and controversies that science and technology present to humanity and the environment. The UN's goals of working together to emit edicts of this nature, to document ethics and human rights, and to guarantee respect for human life are laudable. This concrete common ground in bioethics is a keystone to ensure that laws will eventually recognize this interrelationship.

The goal is that every person can benefit from scientific and technological progress with a respect for his or her fundamental rights and liberties.

The concepts of equality, justice, and respect within human rights have incalculable values and innumerable implications. Human rights in bioethics help us to act ethically and responsibly to ensure the flourishing of these concepts. Bioethics brings together various disciplines but always giving special attention to the right of health protection. This is because a series of needs converge in healthcare—adequate healthcare ensures that no one finds him or herself in a vulnerable or disadvantaged position.

In full keeping with Article 14 (Social Responsibility and Health) of UNESCO's Declaration on Bioethics and Human Rights, the importance of cultural diversity and plurality must always be considered.

18.2 Judaism

Judaism can be thought of as being simultaneously a religion, a nationality and a culture. Of course, Judaism is essentially a religion. But this religion forms the central element of the Jewish culture that binds Jews together as a nation.

In Judaism, the essence of ethics is expressed in its totality in the *Bible*, the ideological base that nurtures the Jewish spirit. The basic cannon of Jewish ethics is based on the teaching of "love thy neighbor as yourself" mentioned in Leviticus 19:18. One of the supreme values of the Jewish ethical system is that of *Guemilut Hasadim*, which in Hebrew means "the giving of loving-kindness." It is quite possible that the substance of this concept rests on the fact that it is a form of assistance that goes beyond charity. It is help that satisfies not only material wants but also spiritual ones.

Every human is expected to fulfil the requirement of good actions toward his or her peers, as this is indispensable for human societal interactions. The first structured

Oral Law, also known as *Mishnah*, is a text that serves to establish jurisprudence. Afterward, rabbinic discussions were transcribed and codified into the *Guemara*. The result of the union of the *Mishnah* and *Guemara* is the *Talmud*, which, along with the *Torah*, forms the base for the development of the *Halakhah*. *Halakhah* is Jewish law, laws or jurisprudence, as mentioned previously by the Halevys.

Judaism has a very rich tradition in medical ethics. Some important aspects of human life that are regulated by Jewish religion are life and death, sexuality, pain, and suffering due to illness.

18.3 Israel Healthcare Model

I would like to respond to the Halevys' paper by stating that I am in complete agreement with them.

Israel's high standards of health services, top-quality medical resources and research, modern hospital facilities, and an impressive ratio of physicians and specialists to population are reflected in the country's low infant mortality rate (3.6 per 1000 live births in 2010) and long life expectancy (81.5 years, average in 2010). Healthcare for all, from infancy to old age, is ensured by the law, and the national expenditure on health compares favorably with that of other developed countries.

The country's population is served by an extensive medical network comprising hospitals, outpatient clinics, and centers for preventive medicine and rehabilitation. Hospital care includes highly advanced procedures and techniques, from in vitro fertilization, MRI scans, and complicated brain surgery to bone marrow and organ transplants.

Mother-and-child care centers, for women during pregnancy and children from birth to early childhood, offer prenatal examinations, early detection of mental and physical handicaps, immunizations, regular pediatric checkups, and health education.

The Israeli healthcare system is characterized by its economic efficiency that stems, among other reasons, from the small number of large HMOs, government control over prices, and strict regulation. Citizens in Israel must have at least basic insurance coverage at one of the country's four HMOs, with 80% of the population belonging to the two largest, offering up-to-date services with modern medical technology.

Israel's health system has been ranked fourth in the world in terms of efficiency, according to the business/finance information company Bloomberg, and fifth out of the 36 member countries in the OECD's Good Living Index.

Despite Israel's commitment to providing health services for all its citizens, by the early 1990s, some six percent of Israelis were not insured through one of the four existing health funds: (1) Kupat Holim Clalit, (2) Maccabi, (3) Beyahad Me'uhedet, and (4) Le'umit. In 1994, the National Health Insurance Law was enacted and it was implemented the following year, rectifying this situation. Since then, all citizens have their health insurance paid by a tax on income (up to 4.8%), while their employer's portion is collected by the National Insurance Institute and passed on to the health insurance fund of the individual's choice.

Israel's comprehensive welfare system is based on legislation which provides for a broad range of national and community services. Care of the elderly; support programs for single parents, children, and youth; prevention and treatment of substance abuse; and assistance for new immigrants comprise a large part of available social services. Correctional services encompass probation frameworks, remedial programs for school dropouts, and residential and observational services for youth in distress. Sheltered workshops and employment counseling are among the rehabilitation services provided for the blind and physically disabled. The developmentally challenged are cared for through various residential- and community-based programs.

18.4 Conclusion

Israel's legal code is secular but incorporates principles and perspective of Jewish law, especially in the areas of danger to life and the provision of healthcare to all.

What I have endeavored in this response was to show how religion, especially the Jewish faith, can be very relevant to the contemporary cultural pluralism in Israel. These are not in opposition. In contrast, religion serves as a guide, orientation, and inspiration for conduct in modern society and protection of human rights.

The state of Israel is a clear leader in the democratic process and protection of human rights. With only a short trip to Israel, one can observe this cultural diversity. Israeli citizens who are relatives of mine and have lived in Israel many years once told me, "We have security and rights to health protection in Israel, even though as in any other system of social security it is limited. But in general, it covers practically everything that we need. An important aspect of the system is that it offers us healthcare without necessarily having to be a taxpayer or employee, and it even includes treatments for cancer and transplants."

Chapter 19
Life in Community: A Christian Response to Jonathan and Adina Halevy

Martha Tarasco

Abstract The *Universal Declaration on Bioethics and Human Rights* includes social responsibility (SR). As a community to which human belongs, it is considered by extension, even if human rights are always individual. SR according to the Talmud means providing in the most opportune moment, and one of the fundamental teachings is co-responsibility. Article 12 states that cultural diversity should not be used at convenience or arbitrarily to limit the application of human rights or their scope. There exists an unstable balance between the individual and the community. The Judaic example of this is circumcision, which is analyzed in this paper from two ethical principles: proportionality and subsidiarity. Judaism has a principle of SR, called *Tikun-holam*, which is to improve the place in which one lives. Without doubt, it is admirable that being a non-proselytist religion, Judaism cares about the lives of others and their environment. Present-day problems of bioethics do not have only *one* single answer in Judaism, as there is no one central authority – akin to the magisterium in Catholicism – which leaves a greater margin for case-by-case interpretations and can result in very different conclusions. In bioethics, SR implies seeing the whole and its parts. When we look forward for SR, we must consider the human dimension integrally and see how the particular elements interact because addition of all the elements does not necessarily arrive at the right answer for an individual or a community. Jewish religious practices have acknowledged the Other and recognized him or her not only as a support but as someone who cannot be fully human without community life. The question regarding practical paradoxes in the application of the Jewish law is something that will require further study and research.

Keywords Social responsibility · Proportionality · Subsidiarity · Cultural diversity

---

M. Tarasco (✉)
Faculty of Bioethics, Anahuac University, Mexico City, Mexico
e-mail: mtarasco@anahuac.mx

J. Tham et al. (eds.), *Religious Perspectives on Social Responsibility in Health*, Advancing Global Bioethics 9, https://doi.org/10.1007/978-3-319-71849-1_19

19.1 Introduction

Social responsibility is a shared goal. In other words, there is no social responsibility with the exclusion of the Other. In this sense, I believe there is a common ethical base that must be considered.[1] The first section of this paper will cover the notion of "social responsibility" from the perspective of the *Universal Declaration on Bioethics and Human Rights*. We will analyze the dilemma between individual and society (liberty/equality) which generates the use of the said term. With an example based in Judaism, we will see how a religious belief can challenge this concept.

In the second section, we see how diverse religions can place different emphasis on this common base. Judaism, like any other religion, has its normativity which helps create what we know as social responsibility. Here we see that rabbinic texts and the Torah are interwoven to provide guidelines toward responsible action.

Finally, we will point out the common paths and lines of thought that exists between Judaism and the Declaration.

19.2 Social Responsibility

The *UNESCO Universal Declaration on Bioethics and Human Rights* gives us a good clue to analyze this notion of social responsibility in bioethics:

> Article 2: Objectives
>
> The objectives of the present Declaration are…to provide a universal framework of principals and procedures that will serve as a guide to the States for the formulation of laws, policies and other instruments in the sphere of bioethics….
>
> f) to promote an equal access to the advances in medicine, science, and technology as well as its most ample circulation possible and a rapid and shared *use of the knowledge relative to these advances and their corresponding benefits, paying equal attention to the necessities of developing countries.*[2]

In the Declaration, the principle of human dignity is not questioned or discussed. Human dignity is affirmed without discussing how it is derived. Nonetheless, from an operative perspective, certain conditions (economic, social, health, etc.) are necessary so that a person can carry out a properly dignified life. In this aspect,

---

[1] I'm not alluding to the minimum ethics doctrine as can be suggested by Adela Cortina: "It makes full sense that a democratic and pluralist society does not want to promote onto its youth, an image of a man admittedly ideal only by some groups that conform it, but it also does not want to resign to transmit to them certain attitudes without which democratic co-existence would be impossible. However, the solution does not consist in changing the 'moral' sign for that of «ethics» in the present assignment, instead we must explain the moral minimums that a democratic society must transmit since we have learned throughout history that these are principles, values, attitudes, and habits to which we cannot resign, without at the same time resigning to our own humanity" (Cortina 2000).

[2] *Declaration.* In: http://portal.unesco.org/es/ev.php-URL_ID=31058&URL_DO=DO_TOPIC&URL_SECTION=201.html

human rights are always individual. Due to the social nature of human beings, however, the community to which he belongs is also considered by extension, without the risk of substantiating the term. Hence Art. 2g of Declaration includes the task "to safeguard and promote the interests of present and future generations."[3]

This is to say there exists an unstable balance between the individual and community. This balance comes to light a bit further on, when it states, "The interest and wellbeing of a person should take precedence with respect to the *exclusive* interest of science or society."

The term *exclusive* gives a clue. Personal interests are not eliminated but cannot exclude those of the community. With these elements we can offer a definition of "social responsibility" from the Declaration as:

> (D1) Those actions that provide the possibility to carry out and/or amplify properly human acts towards their present or future interest which can never be surrendered to the individual or the community.

This constitutes a challenge to the pluralistic society. Article 12 states that cultural diversity cannot be used at convenience or arbitrarily to limit the application of human rights or their scope. Thus, do we not risk of falling into the mere formalization of good intentions?

D1 can be contrasted by other elements that are stated in the Declaration. Article 14 on Social responsibility and health:

1. The promotion of health and social development for its people is an essential task of the governments that is shared by all the sectors in a society.
2. Keeping in mind that the fulfillment of the maximum grade of health which can be achieved is one of the fundamental rights of all human beings without distinction of race, religion, politics, ideology, or social condition, the progress made in science and technology must promote:

    (a) The access to quality medical attention and to the essential medicine, especially for the health of woman and children, since health is essential to life itself and must be considered a social and human necessity
    (b) Access to adequate food and water
    (c) The improvement of life conditions and environment
    (d) The suppression of marginalization and exclusion of persons for any cause
    (e) The reduction *of poverty and illiteracy*[4]

The proposed definition (D1) points out in positive terms what must be done, but the last two items of this article place the emphasis on the elimination of obstacles to human development. With these elements, the definition can be improved:

> (D2) Actions which create the conditions for the possible deployment and/or amplification of the properly human acts, while eliminating also the obstacles regarding their present or future interests which have never been surrendered to the individual or the community.

---

[3] *Ibid.*

[4] *Ibid.*

How do we decide between the situations that fall under the scope of bioethics, in which community and personal interests intersect, but which seem to clash with the principle of human dignity (as in the case of religions)?

We may use the example of male circumcision within Judaism.[5] This practice was prescribed by God in the Old Testament to the descendants of Abraham (Genesis 17). Nevertheless, circumcision in minors who cannot by themselves decide is considered a punishable act in some countries.[6] Circumcision is an example of a religious practice which unites an individual with his community. It is the parents' responsibility with the community and historically speaking, that of communitarian identification.

The UNESCO Declaration additionally points out two ethical principles which are assumed within the concept of human dignity. They are the principles of proportionality and subsidiarity. The principle of proportionality states that one must not attempt good acts that result in disproportionate negative effects. So, there may be actions that – while they may be well-intended and socially beneficial – are illicit. The principle of subsidiarity states that one must help an individual or a community in so far as it is necessary for them to carry out their own functions, but one should not intervene more than it is necessary for the accomplishment of the ends of the community. Now then, does circumcision covers these requisites? Is the religious affiliating act of circumcision a disproportional act or not? Would condemning this practice be an attempt against the autonomy of religion? Even if it is a practice with certain medical risks, is it a religious function in which one must not intervene?

If this were the case, then the individual good would be sacrificed for the community. We can confront this by recognizing that even when these acts (of mutilation) at first glance may affect the individual, the individual benefits from an integration to the community without losing essential physical functions. This type of symbiotic action positively affects a determined person and the community. Even though it is not a therapeutic action but a preventative one (evidence-based medicine has shown that circumcision is an adequate hygienic measure), its meaning for Judaism is quite diverse.[7]

After presenting the notion of social responsibility and offering an example in Judaism, we may consider the following: In what way can Judaism enrich the reviewed definition of social responsibility? What are its contributions to the notion of social responsibility?

---

[5] This is a practice that is found in ancient civilizations that predates the practice in Mesapotamia. See (Bigelli and César 2006).

[6] See (Silva Sánchez 2013, p. 12).

[7] While there is recent evidence that circumcised males have reduced risk of contracting HIV and other STDs, this is not the main reason for its practice in the Jewish religion. See (Bamigboye and Igberase 2010).

19.3 Judaism and Social Responsibility

In Article 14c of the Declaration, we read that social responsibility and health include "the improvement of life conditions and environment."

Judaism has a principle of social responsibility that precedes the Declaration or the use of the term social responsibility. This is the principle of *Tikun-holam* which is to improve the place in which one lives. It contains many other aspects, with differences between orthodox and secular Judaism. Without doubt, it is admirable that being a non-proselytist religion, Judaism cares about the lives of others and their environment.

The present work does not pretend to analyze "all the Jewish sources." In Judaism, there is a written source of the *Torah* and an oral in nature called the *Talmud* (Tapia-Adler 2010). In the latter, one needs to make an exegesis to better interpret the rule according to the situation. Present-day problems of bioethics, like abortion, do not have only *one* single answer in Judaism (Silber 1980). In Judaism there is no one central authority – akin to the magisterium in Catholicism – which leaves a greater margin for case-by-case interpretations and can result in very different conclusions (Tapia-Adler 2010). Rabbi Byron Sherwin gives such an example in his book *Why be Good?*:

> It is told that one day the rabbi Rhyzen surprised some of his disciples playing checkers. Noting his presence, the students were ashamed of losing time in leisure instead of studying the sacred texts. However, he said: I'm happy that you play checkers because its rules are also the rules of spiritual and moral development. First: move pieces from square to square without skipping any of them. Second: always move forward, never go back. Third: once you reach one of the squares of the last row, the piece – turned into a queen – can go back, jump a number of squares, and move in any direction. (Sherwin 1999)

This text illustrates the legalistic character that Judaism tends to avoid deontology. Rules by themselves do not guarantee acting morally. The metaphor also illustrates that prudence, without the exercise of rules, would fall into consequentialism. Using this parable and its three rules, we can re-elaborate the concept of social responsibility this way:

1. Social responsibility in bioethics is accomplished step by step. It is not possible and would be naïve to think that there is a solution in a linear sense of a problem. It is in a technical sense a complex solution.[8] The complexity is caused by the reading of reality which consists in balancing different components at different

[8] "Complex systems are constituted by heterogeneous elements in interaction – hence their denomination as complex – which means that their subsystems belong to the material domains of very diverse disciplines… This does not mean, however, that it is easy to overcome the practical difficulties of articulating such studies. Interdisciplinary supposes the integration of different disciplinary perspectives, by which it is necessary that each one of the members of a research team be an expert in his own field. In this sense, the research team is multidisciplinary. The fundamental difference between an interdisciplinary investigation and the so called multi- (or trans-) disciplinary is in the manner of conceiving a problem and in the common denominator that the members of the investigation team share." (García 2006).

levels. Social responsibility implies seeing the whole and its parts simultaneously. The checkers metaphor reminds us that the rules of the system permit a certain balance when we interact with the elements. In a similar manner, when we look for "social responsibility," we must consider the human dimension integrally (which we call *dignity*) and see how the particular elements (health, housing, education, etc.) interact. In other words, we do not necessarily arrive at the right answer for an individual or a community by adding up all the elements as an equation. Certain interaction of the elements in a determined situation will generate a correct balance, but this cannot be applied in every system.[9]

2. The "nonstop metaphor." The interpretation of human reality is continuous both in human development and in the Torah of Judaism. There cannot be a definitive reading that solves every ethical situation. There are specific answers for given circumstances. The answer must be found in each situation; *but* this does not mean that it is totally arbitrary.[10]
3. The third element is "prudence." When the rules are understood and followed and virtuous goals are reached, the subject of action – with improved habits – will arrive to a point of "stepping back or advancing." What seems like an act of courage may in fact be an irresponsible action, and so one should take a step back. On the contrary, what at first glance seems prudent can be a cowardly action. Prudence belongs to the realm of traditional ethics (Aristóteles 1871) and is not only Jewish.

What elements of judgment and action are useful in the Jewish tradition? I think that the answer is the golden rule: "You will not seek revenge nor maintain anger towards the sons of your people, instead you will love your neighbor as you love yourself; I am the Lord."[11] This golden rule, common to other religions, can take on a more concrete norm in the Babylonian *Talmud*:

> Anyone who can prevent a sin, and does not, will be himself considered responsible, as he could have dissuaded the sinner…. What preceded is applied if one esteems that there exists a possibility of success, but if one knows that the sinner will not pay absolutely any heed to his observations, then he must not intercede, since, in the same manner in which it is his duty to intervene when he will be listened to, it is recommended that he abstain from this in the opposite case. (Talmud 2003)

---

[9] "While in the case of the multidisciplinary investigations they usually imply a summation of the collaborations of each of the investigators carried out from the scope of their own discipline with regard to a general problem that can be analyzed from different perspectives; an interdisciplinary investigation supposes the integration of these different scopes for (that is to say, before) the delimitation of a problem. To say it in a different manner, while in one case what is integrated are the results of different studies based on a common problem, in the case of an interdisciplinary investigation the integration of the different perspectives constitutes the delimitation of the problem." (García 2006).

[10] For a greater analysis of the interpretation of Scripture between the Old and the New Testaments, (Bosca 2008).

[11] *Levíticus*, 19, 18.

One of the fundamental teachings in Judaism is co-responsibility. As noted in the principal of proportionality, one must not attempt good acts that may result in disproportionate negative effects. Clearly, there is much debate on what is considered good. Nevertheless, social responsibility according to the Talmud means providing correction in the most opportune moment. There is another text that complements it as a negative expression of the principle or as a limited golden rule:

> It is forbidden to humiliate your neighbor, both with words or with actions, and even more so, in public. So our Wise Men said: "He who shames his neighbor in public will have no part in the future world" (Pirké Avot 4). From the latter, one must take great care in not humiliating in public any person, child or adult, or calling him by a nickname which can embarrass him or by telling in his presence any story that may disturb him.[12]

The advice is to not intervene to harm or act to damage. Both interpretations of the Talmudic text remit us once again to the indicated formal principle. Both these formulations are essentially reducible to the principle of "do good and avoid evil." (Aquinus 2001)

19.4 Final Remarks

The Declarations on human rights have acknowledged the interdependency of all human beings in striving for the common good. We need minimum common rules to promote human development and flourishing. The *Universal Declaration on Bioethics and Human Rights* is a deeper acknowledgment of the limits of human development. It is a collective effort with a shared axiological background in which the ends are more than expression of desires. We visualize not only the motives for action, which are infinite, but also the reasons for them. This is the meaning of being an "independent ethical reasoner." But "independence" does not imply mere subjectivity or isolation.

Jewish religious practices have acknowledged the Other and recognized him or her not only as a support but as someone who cannot be fully human without community life. The question regarding practical paradoxes in the application of the Jewish law is something that will require further study and research.

The words of the Jewish thinker, Hannah Arendt, best explain the notion of social responsibility as a way of practicing politics in the most classical sense of the word – in the life of the community.

The need to understand what has happened is an act of responsibility by which we comply our human obligation to take care of the world. On one hand, it is breaking with the tradition. And on the other, it is an ineludible political responsibility by the sole act of being part of humanity.[13]

---

[12] Talmud, *Babá Metsiá* 59; Rambam, Deot 6/8.

[13] (Novo 2008) Departamento de Filosofía, Facultad de Humanidades y Ciencias de la Educación Universidad Nacional de La Plata 2008, p. 1.

References

AAVV. Circuncisión. Quick Reference, *Master Catalog of practical clinic guides: SSA-289-10,* Consejo de Salubridad General, México, http://www.cenetec.salud.gob.mx/descargas/gpc/CatalogoMaestro/289_SSA_10_Circuncision/GRR_SSA_289_10.pdf.

Aristóteles. 1871. *Ética a Nicómaco*. Madrid: Espasa Calpe.

Aquinus, Thomas. 2001. *Summa Theologiae*, C. 79, a. 12, c. BAC, Madrid, p. 738.

Bamigboye A.A., and G.O. Igberase. 2010. *Circuncisión masculina para la prevención de la adquisición del VIH en hombres heterosexuales: Commented by BSR* (last revision: January 1, 2010). Reproductive health library OMS; Geneva: World Health Organization. http://apps.who.int/rhl/hiv_aids/Cd003362_igberasego_com/es/. Consulted on August 12, 2014.

Bigelli, Potenziani, César, Julio. 2006. Historia de la circuncisión y su trascendencia en las diferentes culturas de la humanidad In *Temas Actuales en Urología*, 249–286. Moore de Venezuela SA.

Bosca, R. 2008. *"Caritas y tzedaká. "Caritas y tzedaká.* las fuentes veterotestamentarias de la doctrina social de la iglesia en el marco del diálogo hebraico-católico. In *Scripta Theologica*, Universidad de Navarra, vol. XL/1.

Caro, Y. 2003. *Shul'han Aruj* (compilation), Hasdé Lea foundation. México: Editorial Jerusalem de México, México.

Cortina, A. 2000. *Ética Mínima*. Madrid: Tecnos.

De Aquino, T. 2001. *Suma de Teología*. Madrid: Biblioteca de Autores Cristianos.

García, R. 2006. *Sistemas complejos*. Barcelona: Gedisa editorial.

Gómez Álvarez, J.E. 2006. *Derechos Humanos, bioética y religión: recuperación del florecimiento humano*. México: Comisión de Derechos Humanos del Estado de México.

Neusner, J. 2008. *The golden rule in classical Judaism*. Bard College.

Novo, M. Rita. 2008. El concepto de, "Responsabilidad" en la filosofía política de Hannah Arendt. *Actas de las VII Jornadas de Investigación en Filosofía.*

Potenziani Bigelli, J.C. 2006. Historia de la circuncisión. In: *Temas actuales en urología*, Moore de Venezuela SA, 249–285. http://www.sexarchive.info/BIB/Potenziani/hist02.htm.

Sherwin, B.L. 1999. *¿Por qué ser bueno?* Barcelona: Plaza y Janés.

Silber, T. 1980. Abortion: A Jewish view. *Journal of Religion and Health* 19 (3): 231.

Silva Sánchez, J.M. 2013. Circuncisión infantil. In: Dret Penal, 1.

Talmud. 2003. *Yevamot* 65b, taken from Caro, Yosef. *Shul'han Aruj* (compilation). México: Hasdé Lea foundation, editorial Jerusalem de México.

Tapia-Adler, A.M. 2010. Notes on Judaism and bioethics. *Bioethics Act* 16: 25–26.

Part VII
Conclusion

Chapter 20
Convergence and Cooperation in Social Responsibility in Health

Alberto García Gómez and Claudia Ruiz Sotomayor

Abstract Global health has been defined as the area of study, research, and practice that places a priority on improving health and achieving equity in health for all people worldwide. Despite the efforts of achieving equity, global health is characterized by disparities that have led to some discussion about who is responsible for providing healthcare. The UNESCO Chair of Bioethics and Human Rights has provided a necessary space for reflection to address the dilemmas of resource allocation, global health, and social responsibility using Article 14 as a guideline and starting from a global bioethics framework. This essay explores the relationship between global health and social responsibility and their link to global bioethics, arguing that the approach of global bioethics based on rights and responsibilities is an important tool to achieve convergence and cooperation regarding social responsibility and health, because it offers proper venues for dialogue and understanding.

Keywords Global health · Global bioethics · Social responsibility · Convergence · Cooperation

20.1 Introduction

Global health conditions are marked by inequities due mostly to poverty and lack of access to healthcare services. It is known that life expectancy is shorter and preventable diseases are not well addressed in poor communities. People with no access or limited access to healthcare tend to have more complications of treatable diseases (Wilkinson and Marmot 2003). These global health disparities have led to

A. García Gómez (✉)
UNESCO Chair in Bioethics and Human Rights, Rome, Italy
e-mail: agarcia@unescobiochair.org

C. Ruiz Sotomayor
MD Anderson Cancer Center, Houston, TX, USA
e-mail: csotomayor@unescobiochair.org

J. Tham et al. (eds.), *Religious Perspectives on Social Responsibility in Health*, Advancing Global Bioethics 9, https://doi.org/10.1007/978-3-319-71849-1_20

some discussion about who is responsible for providing healthcare. The international community supports the idea that there is a moral duty held by all societies to promote health, to prevent and treat diseases, and to provide the highest attainable standard of health. Article 14 of the UNESCO Universal Declaration on Bioethics and Human Rights (2005) introduces the principle of social responsibility and health in the field of bioethics "to ensure, whenever possible, that progress in science and technology contributes to justice, equity and to the interest of humanity." The stakeholders are numerous and include governments, groups of people organized within societies, commercial companies, political organizations, educational institutions, and others.

The UNESCO Chair of Bioethics and Human Rights has provided a necessary space for reflection to address the dilemmas of resource allocation, global health, and social responsibility using Article 14 as a guideline. The approach suggested to focus in a global bioethics framework that emerged as a result of the increasing awareness of the interrelatedness of individuals and their ethical dilemmas and the recognition that global problems needed global solutions (Tham et al. 2017). Global bioethics is developing as a nexus between global health and biopolitics at an international level, not only as an academic approach but one supported by social leaders and governments.

This essay explores the relationship between global health and social responsibility and their link to global bioethics, arguing that the approach of global bioethics based on rights and responsibilities is an important tool to achieve convergence and cooperation regarding social responsibility and health, because it offers proper venues for dialogue and understanding.

20.2 Global Health and Social Responsibility

It is known that there are large inequalities in health in the world. By the year 2003, the life expectancy in Japan is 80 years; in Malawi, 40 years. The under-five mortality rate in Norway is 4/1000; in Sierra Leone, 316/1000. The maternal mortality rate in Sweden is 5/100000; in Mozambique, 1000/100000 (Dwyer 2003). Thanks to the United Nations Millennium Development Goals (MDGs), there is progress in child survival worldwide, and by 2013 child deaths have almost halved – falling from an estimated 90 deaths per 1000 live births to 46 deaths per 1000 live births. However, despite great advances, this is not enough to reach the goal of reducing the death rate by two-thirds. Less than one-third of all countries have achieved or are on track to meet this target by the end of 2015. According to the World Health Organization (WHO), in the year 2015, life expectancy at birth has increased 6 years for both men and women since 1990, but in low- and middle-income countries, only two-thirds of pregnant women with HIV receive antiretrovirals to prevent transmission to their baby. Only 1 in 3 African children with suspected pneumonia receives antibiotics. The median age of people living in low-income countries is 20 years, while it is 40 years in high-income countries (WHO 2015).

Given such disparities, it is necessary to have a global understanding of health and to create local solutions. Global health has been defined as the area of study, research, and practice that places a priority on improving health and achieving equity in health for all people worldwide (Brown et al. 2006). Global health draws on the philosophy of health and human rights (Pinto et al. 2009), which is based on the inherent value of each person and the claims one has on the local and global community. It is clear then that there are many reasons to support the promotion of health and the provision of healthcare globally. First, it protects each individual's human dignity and personal integrity; second, better health allows for greater participation in the community; and third, by providing basic primary care, many complications can be avoided, reducing fatalities and the total cost of healthcare. But the remaining question is: Who is responsible for such provisions?

On a global level, social responsibility in healthcare refers to a moral duty held by all societies to promote health, to prevent and treat diseases, and to provide the highest attainable standard of health.[1] Since the United Nations Universal Declaration of Human Rights of 1948, UN documents tend to have a more globalized approach to the responsibility for health. In its Article 25, it is proclaimed that it is a right to have a standard of living that ensures the health and well-being of the person. These goals have been recognized by the international community as a central purpose of governments, and so it has placed a lot of responsibility in each state (UNESCO 2005). However, the state is typically an entity which can only act by law and is bound by special responsibility to its citizens. Allocating state money for international aid could be considered by some states, but for others it is difficult to allocate resources to foreigners when there are many citizens lacking access to decent medical treatment in their own country. It would be contrary to the idea of state responsibility not to give them priority before assisting foreign countries.

It can be concluded then that the social responsibility of each state is toward its own country, but even then this view on social responsibility causes friction among citizens, for example, libertarians assert that the individual should be free to decide how to use his/her personal means to meet his/her various needs (Menzel 2009). Some argue that individuals should care for their own health and help pay for their own healthcare (Resnik 2007). Neoliberalists argue that the state should avoid being involved in the direct provision of healthcare and leave most of it to private practice because every individual should be made responsible for his/her own health and healthcare. Others would argue that state involvement would invariably result in the increase of taxes, creating a tax burden upon society, and an unjust disposition of each person's income (Taylor and Hawley 2010).

Egalitarians reject these libertarian views arguing that when supporting individual responsibility, people on the bottom of the economic ladder (Sachs 2005) are marginalized because they do not have the necessary means to pay for their healthcare. In this era there are many individuals and communities that are especially

[1] Highest attainable standard of health: An effective and integrated health system, encompassing medical care and the underlying determinants of health, which is responsive to national and local priorities and accessible to all.

vulnerable due to structural violence (Farmer 2003). Romer (1988) argues that in order to assure equality of opportunity in health, social intervention is necessary because "health is a pre-requisite for successful achievement, illness or injury in any person is unpredictable, and obtaining medical care and a healthful environment is beyond the resources of the individual and requires collective action." This argument goes with the line of what Menzel (2009) has called a "Just Sharing Principle." In Menzel's version of this principle, the financial burdens of "medical misfortunes" should be shared equally between healthy and unhealthy people, unless individuals can or have control over these misfortunes (Kelleher 2014).

The *Report of the International Bioethics Committee of UNESCO on the Principle of Respect for Human Vulnerability and Personal Integrity* (2013) states that social vulnerability is caused or exacerbated by a lack of means and of the capacity to protect oneself. In many developing countries, the lack of means and access to proper healthcare affects millions of people, making them vulnerable. The lack of capacity to protect oneself is a cause for the community, agencies, and governments to step up and provide the necessary means to help people thrive and be healthy (Tham et al. 2014). But the remaining questions are how can we translate theory into practice? How can we create conscience in our health authorities? Given that the needs of each community are different, what care should be included in a just and basic minimum?

20.3 Social Justice in a Multicultural World

There are different theories of distributive justice that can be used to try to give a systematic approach to the dilemma of distributive justice and resource allocation in healthcare around the globe. The meritocratic interpretation of distributive justice was first seen in Plato's *Republic* where natural ability was the basis of the three classes: guardians (rulers), auxiliaries (warriors), and producers (farmers, artisans). This interpretation holds that all persons should be treated according to their merits, achievements, and abilities (Daniels 1978). It is understood then that it is required that persons have merits, abilities, etc., in order for the ideal of justice to be operational; however, for some, this cannot be applied to healthcare because a meritocratic view encourages discriminations among those who have merits and those who do not, which cannot be applied in basic needs for human beings, such as health and healthcare.

In *A Theory of Justice* (1999), Rawls argues that the way to think about justice is to ask what principles we would agree to in an initial situation of equality. For example, Rawls assumes that some natural goods are distributed unequally in a random way and that while this distribution can be influenced by social policies, it is not directly under social control. Since individuals have no control over their own natural goods, these goods are undeserved, and since these goods are undeserved, they are arbitrary from a moral point of view. Thus, one of the main tasks of social justice is to ensure that people still have fair equality of opportunity despite their natural inequalities, and this constraint implies principles of redress to compensate for natural inequalities (Resnik 2007).

Daniels, inspired by Rawls but claiming that health and healthcare should be treated different, argued that the problem of resource allocation in healthcare with what he calls the "accountability for reasonableness" is an attempt to connect views about deliberative democracy to decision-making at various institutional levels, whether public or private, in the complex health systems around the globe. To achieve this accountability, he proposes to place great emphasis on the importance of publicity as a constraint on theories of justice. According to Daniels, accountability for reasonableness pushes decision-makers toward finding reasons all can agree are relevant to the goals of cooperative health-delivery schemes. In this way, accountability for reasonableness promotes the democratic deliberation that Rawls also advocates (Rhodes et al. 2002).

This accountability of reasonableness proposed by Daniels has to be linked to the virtue of justice so a true democratic deliberation where a fair distribution of healthcare can be achieved. The idea of justice is first found in book I of Plato's Republic: *Justice consists on rendering to each his due* (Thompson 2014). For Aristotle, justice consists in what is lawful and fair, with fairness involving equitable distributions and the correction of what is inequitable (Pomerleau 1995). Situations and communities are just, according to Aristotle, when individuals receive benefits according to their merits, or virtue: those most virtuous should receive more of whatever goods society is in a position to distribute (exemptions from various burdens or evils counting as goods). Aristotle treats the virtue of individual justice as a matter of being disposed to properly respect and promote just social arrangements (Slote 2014). Plato and Aristotle influenced medieval philosophers; however Christian writers were downplaying another central element in Christian thought and morality, the emphasis on agapic love. For Augustine, the cardinal virtue of justice requires that we try to give all people their due; for Aquinas, justice is that rational mean between opposite sorts of injustice, involving proportional distributions and reciprocal transactions (Pomerleau 1995); he argued that "the proper act of justice is nothing else than to render to each one his own" *(Summa Theologica 2:57).*

It can be said then that the notion of justice is contingent upon the existence of a moral standard (Hodge 2012). The ability to determine whether or not an action is just or unjust requires the presence of an external standard through which an action is evaluated. In order to identify injustice, it is necessary to have an independent moral framework for evaluating the world (Hodge 2012).

But, who determines that standard? Where does the world get a moral compass?

For many years religions have played an important role in developing such a standard. By positing moral values, spiritual traditions provide an external framework for assessing the world (Hodge 2012). One way to explore the possibility of a relatively universal system of norms is through the notion of religion. Knowing that nearly 90% of the world's population identifies with a religion, then it is worthwhile to find whether they have similarities in the way they approach to justice and then in regulating the life sciences (Tham et al. 2017).

The richness of our multicultural society is in part because of the moral standards that all different religions are positing. When analyzing social responsibility in healthcare, it is important to consider the input given by these religions/traditions. Their understanding of justice, responsibility, and rights is crucial to develop a full understanding of the needs of the community as well as the fairest way to provide with health and healthcare to our society. Each and every one of the religions offer different perspectives as well as a variety of solutions to a global problem.

For Buddhism, the moral standard of justice and social responsibility rely on their views of active compassion shown in the Buddhist doctrines of *dāna* (gift-giving, practice of generosity) and *punya* (doing good). Confucians and their sense of moral duty to family offer an important family-based and family-oriented ethics, where special regard is given to women and children. Islamic views of God and their duty toward him bring a new perspective on human rights and responsibilities, thanks to their concepts of the "rights" of mankind, *huquq al-ibad*, and communal obligation, *fardh kifayah*. Hinduism views on transcendence and selfless actions promote social welfare as a part of spiritual discipline that leads to an intrinsic desire of men to help others. Judaism argues for universal equality based on the assumption that every individual has the right to expect that society will meet all his or her particular needs in full. Christianity offers a view of universal love based in the parable of the Good Samaritan (Luke 10: 25–37), where Jesus opened a new way of seeing the neighbor and a new way of caring for others by giving more than is expected and introduced the technical concept of *supererogation* as referring to actions that go beyond the call of duty. Works of supererogation are those which it is virtuous to do but not absolutely required. They express the difference between the Old Law (of duty) and the New Law (of love) (Heyd 1982).

20.4 Human Rights and Responsibilities in the Healthcare Discourse

Most of the discussion in bioethics concerning access to healthcare has focused on the difficulty of recognizing the right to healthcare and the role of the state in the provision of such right. Some authors (Baumrin 2002) critique the idea of a right to healthcare and state that there is not such a right but rather that it is an *incoherent rhetorical flourish.* For some cultures and traditions, the recognition of healthcare as a right is problematic because it is their belief that to have a right there must be a provider, meaning someone responsible for the provision of such a right, and since in some states there is no acceptance of such responsibility, therefore there would be no right. The right that establishes *an ought* is contingent on some agent to accomplish the task, and if the agent is not capable of fulfilling that task, then the duty is placed on hold. Ultimately, if the duty is not fulfilled, it is not genuine, and if there is no genuine duty, there is no genuine right (Baumrin 2002).

On the contrary, Hessler and Buchanan (2002) argue that there is a right to healthcare because there is a global need to protect the basic human interest in health. These authors reason that by acknowledging a right to healthcare, the state and other international agencies would be held accountable for the provision of healthcare. The state would be fostering equality or at least compensating for some of the inequalities. The fact that some rights are not always fulfilled does not mean they are not genuine rights; what it means is that more efforts should be made by society to fulfill them. Some authors state that the intention of fulfillment is enough to prove the existence of such right and that all human beings have the duty to provide necessary health assistance, in accordance with their means (Lango 2002). By viewing then these duties as having corresponding rights, there now exist agents to demand their fulfillment and thus put pressure on society to find creative solutions to overcome the limitations of resources.

Even when there is no evidence of getting close to a consensus on whether there is a right to healthcare or not, there are many international declarations stating that governments and international agencies have a social responsibility to provide such care. Let us remember that human rights and responsibilities are born from a human need to address injustice and to protect human dignity, especially of those who need that protection the most. Those persons in vulnerable positions, victims of the poverty trap (Sachs 2005), desperately need aid and protection from deprivation. The duty of providing healthcare rests with society, government, international agencies, religious groups, and others. In short, it is the responsibility of everyone who is in a position to help. If duties are correlated with a right (Rainbolt 2006) and the provision of healthcare is a duty, then it is fair to say that the receipt of healthcare is a right. The access to healthcare and the highest attainable standard of health keeps not only an individual healthy and happy but a community active, functioning, and thriving. Recognizing the right to healthcare is just a first step to a real improvement in resource allocation. By focusing on the duty to provide healthcare, and in turn on the right to receive healthcare, both the providers and the consumers of healthcare will have a voice in the conversation and help to realize a more equitable distribution of healthcare.

20.5 Conclusion

Global health draws on the philosophy of health and human rights, which is based on the inherent human value: dignity. Global health is concerned with fulfilling these claims and seeking a world where all enjoy a certain standard of health and healthcare. At the same time, global bioethics has developed a human rights framework that has been adopted by international agencies such as UNESCO, because it is in their belief that "the human rights framework provides a more useful approach for analyzing and responding to modern public health challenges than any framework thus far available within the biomedical tradition." The Universal Declaration on Bioethics and Human Rights is an important step to achieve global bioethical standards. Its main objective is to assemble some basic standards to help states in their efforts to promote responsible biomedical research and clinical practice, in conformity with the principles of international human rights law (Adorno 2007). This author proposes that the notion of individual and collective views on dignity could be the common ground of all basic rights and freedoms and gives value to humanity (Adorno 2009). The incorporation of local cultures and beliefs is essential to a responsible and compassionate healthcare policy. The UNESCO Declaration inscribes itself in this trend when it places at the top of its principles that of "human dignity, human rights and fundamental freedoms" (Article 3.1).

It is clear that there are many facts to support the promotion and engagement to global health and global bioethics. Global bioethics helps us to protect everyone's human dignity and personal integrity; and global health opens our eyes to create opportunities and better environments to improve the overall health of the population by providing basic primary care. By assuring access to the highest attainable standard of health, many complications can be avoided, fatalities would be reduced, and the total cost of healthcare would be less. The provision of healthcare globally is justified as a social responsibility because humans are not isolated from each other, we are social beings that depend on each other to subsist, and that fact creates a duty toward one another to build a successful community. Global and regional disease eradication efforts (e.g., tuberculosis, polio, and measles) are examples of this commitment as a community to global health equity that requires the cooperation of global, regional, state, and private institution actors in a common cause (Ruger 2012).

In a growing secular view in the world, it seems that the notion of a global understanding of bioethics is becoming a trend that endeavors to cover all possible ethical problems arising with regard to life, health, environment, politics, etc. (Schoedrer 2005). However, the problem that the secular world face when addressing health disparities and ethical dilemmas in a global context is that there are many cultures, religions, and economic circumstances that need different approaches (Hellsten 2008). The probability of facing relativism is evident if there are no standardized guidelines or a common ground. While respecting differences, emphasizing the striking commonality at the heart of most religions can serve as a moral compass. It is imperative the recognition of hermeneutical diversity so there can be an establishment of a set of multiple interpretations of a shared norm. Learning others' moral vocabulary can lead us to an empathetic view of tolerance, without falling into moral relativism.

In a multicultural world, global health has relied their philosophy on the idea of human rights, which is somewhat in conflict with some cultures and belief systems, that is why global bioethics has responded to the dilemma by creating proper forums of dialogue and conversation within cultures and traditions to achieve convergence and common ground. The UNESCO Chair in Bioethics and Human Rights has provided the proper channels of communication by creating a forum of diverse bioethics thought leaders of different religious backgrounds so that this understanding and convergence can be achieved.

References

Andorno, R. 2007. Global bioethics at UNESCO: In defence of the universal declaration on bioethics and human rights. *Journal of Medical Ethics* 33 (3): 150–154.

———. 2009. Human dignity and human rights as a common ground for a global bioethics. *Journal of Medicine and Philosophy.* 34 (3): 223–240. Print.

Baumrin, S.B. 2002. Why there is no right to health care. In *Medicine and social justice: Essays on the distribution of health care*, ed. Rosamond Rhodes, M.P. Battin, and Anita Silvers. Oxford: Oxford University Press. Kindle version.

Brown, T.M., M. Cueto, and E. Fee. 2006. The World Health Organization and the transition from "international" to "global" public health. *American Journal of Public Health* 96: 62–72.

Daniels, N. 1978. Merit and meritocracy. *Philosophy and Public Affairs* 7: 206–223.

Dwyer, J. 2003. Teaching global bioethics. *Bioethics* 17 (5/6): 432.

Farmer, P. 2003. *Pathologies of power: Health, human rights, and the new war on the poor.* Berkeley: University of California Press, Print.

Hellsten, S. 2008. Global bioethics: Utopia or reality? *Developing World Bioethics.* 8 (2): 70–81. Print.

Hessler, K., and A. Buchanan. 2002. Equality, democracy, and the human right to health care. In *Medicine and social justice: Essays on the distribution of health care*, ed. Rosamond Rhodes, M.P. Battin, and Anita Silvers. Oxford: Oxford University Press. Kindle version.

Heyd, D. 1982. *Supererogation: Its status in ethical theory* (Cambridge studies in philosophy). Cambridge [Cambridgeshire]/New York: Cambridge University Press.

Hodge, D. 2012. The conceptual and empirical relationship between spirituality and social justice: Exemplars from diverse faith traditions. *Journal of Religion & Spirituality in Social Work: Social Thought* 31 (1–2): 32–50.

Kelleher, J. 2014. Beneficence, justice, and health care. *Kennedy Institute of Ethics Journal.* 24 (1): 27–49. Print.

Lango, J. 2002. Global health, human rights, and distributive justice. In *Medicine and social justice: Essays on the distribution of health care*, ed. Rosamond Rhodes, M.P. Battin, and Anita Silvers. Oxford: Oxford University Press. Kindle version.

Menzel, P. 2009. Justice and fairness: Mandating universal participation. *Hastings center report* 39 (3): 4–6. Print.

Pinto, A., E. Ross, and G. Upshur. 2009. Global health ethics for students. *Developing World Bioethics* 9 (1): 1–10. Academic Search Complete.

Pomerleau, W. 1995. Western theories of justice. *The internet encyclopedia of philosophy.* Available at: ISSN 2161-0002, http://www.iep.utm.edu/.

Rainbolt, G.W. 2006. Rights theory. *Philosophy Compass* 1 (1): 11–21. Print.

Rawls, J. 1999. *A theory of justice.* Rev. ed. Cambridge, MA: Belknap Press of Harvard University Press.

Report of the International Bioethics Committee [Internet] Paris, France: *The principle for respect of human vulnerability and personal integrity* [2013; cited 2014 Nov. 26]. Available at http://unesdoc.unesco.org/images/0021/002194/219494E.pdf.

Resnik, D. 2007. Responsibility for health: Personal, social, and environmental. *Journal of Medical Ethics* 33 (8): 444.

Rhodes, R., M.P. Battin, and A. Silvers. 2002. *Medicine and social justice: Essays on the distribution of health care.* Oxford: Oxford University Press.

Romer, R. 1988. The right to health care-gains and gaps. *American Journal of Public Health* 78(3): 241–247. Print.

Ruger, J. 2012. Global Health justice and governance. *American Journal of Bioethics* 12 (12): 35–54.

Sachs, J. 2005. *The end of poverty: Economic possibilities for our time.* New York: Penguin Press.

Schoedrer, D. 2005. Human rights and their role in global bioethics. *Cambridge Quarterly of Healthcare Ethics* 14 (2): 221–223. Print.

Slote, M. 2014. "*Justice as a virtue*" *the Stanford encyclopedia of philosophy* (Fall 2014 Edition), ed. Edward N. Zalta. Available at: http://plato.stanford.edu/archives/fall2014/entries/justice-virtue/.

Taylor, G., and H. Hawley. 2010. *Key debates in health care*. Maidenhead: Open University Press. Internet resource.

Tham, J., A. Garcia, and G. Miranda, eds. 2014. *Religious perspectives on human vulnerability in bioethics*. Dordrecht: Springer Advancing global bioethics.

Tham, J., M. Kwan, and A. Garcia, eds. 2017. *Religious perspectives on bioethics and human rights*. Dordrecht: Springer Advancing global bioethics.

Thompson, S. 2014. *Encyclopedia of diversity and social justice*. Retrieved from http://www.eblib.com.

UNESCO.org [internet] Universal declaration of Bioethics and Human Rights [2005, cited 12/01/14]. Available from: http://portal.unesco.org/en/ev.php-URL_ID=31058&URL_DO=DO_TOPIC&URL_SECTION=201.html.

WHO.org [internet] World Health Organization, World Health Statistics reports on global health goals for 194 countries [updated 2015; cited Jan 6, 2016]. Available from: http://www.who.int/mediacentre/news/releases/2015/world-health-statistics-2015/en/.

Wilkinson, R., and M. Marmot. 2003. *Social determinants of health: The solid facts*. Copenhagen: World Health Organization, Regional Office for Europe.

Index

J. Tham et al. (eds.), *Religious Perspectives on Social Responsibility in Health*,
Advancing Global Bioethics 9, https://doi.org/10.1007/978-3-319-71849-1